Monographs on Astronomical Subjects: 9

General Editor, A. J. Meadows, D.Phil.,

Professor of Astronomy, University of Leicester

The Origin and Evolution of Planetary Atmospheres

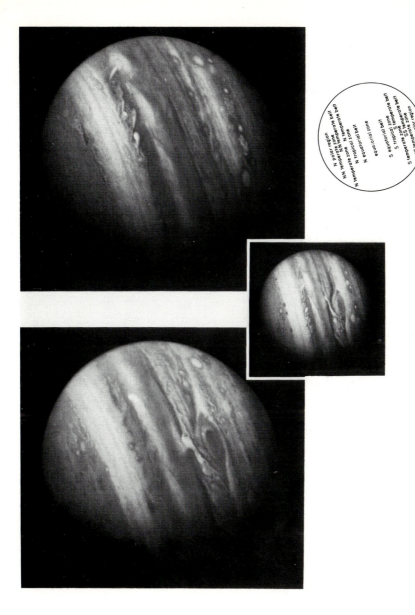

N polar region
N N temperate belt
N temperate zone   N temperate belt
N temperate zone   N
N tropical zone
N equatorial belt
equatorial zone
S equatorial belt
S tropical zone   S temperate belt
S temperate zone   S
S temperate zone   SS temperate belt
SS temperate zone
S polar region

Frontispiece. Voyager photographs of Jupiter. (Photograph courtesy of National Aeronautics and Space Administration, USA.)

Frontispiece. Two photos of Jupiter taken by Voyager 2 contrast sharply with the small inset photo by Voyager 1, taken almost four months earlier. They demonstrate that the planet's atmosphere undergoes constant changes, and that, although individual clouds are long lived, winds blow at greatly different speeds at different latitudes, causing clouds to move independently and pass each other. The small photo was taken on 24 January 1979, by Voyager 1 while 40 million kilometres from Jupiter. The photo on the left was taken on 9 May, when Voyager 2 was 46.3 million kilometres distant. The photo on the right was taken two hours earlier. A Jovian day is about 10 hours long. Comparison of the small image with the large ones shows that many features on Jupiter have moved great distances. A white oval that, in the small photo, is just below and to the left (west) of the Great Red Spot, has moved 60° eastward (right) around the planet in the Voyager 2 images; that allowed the band of clouds just west (left) of the oval to move directly beneath the red spot. Bright clouds near the edge of the dark equatorial band have moved around the planet since January. A bright tongue extending upward from the Great Red Spot is interacting with a thin bright cloud above it that has travelled twice around Jupiter in four months. The Voyager Project is managed and controlled for NASA's Office of Space Science by the Jet Propulsion Laboratory. (The inset shows the nomenclature used to describe the zoned regions on Jupiter.)

The Origin and Evolution of Planetary Atmospheres

A. Henderson-Sellers

University of Liverpool

Monographs on Astronomical Subjects: 9

Adam Hilger Ltd, Bristol

British Library Cataloguing in Publication Data

Henderson-Sellers, A.
The origin and evolution of planetary atmospheres. — (Monographs on astronomical subjects ISSN 0141-1128; 9)
1. Planets—Atmospheres
I. Title   II. Series
523.4   QB603.A85

ISBN 0-85274-385-8

Published by Adam Hilger Ltd, Techno House, Redcliffe Way, Bristol BS1 6NX.

The Adam Hilger book-publishing imprint is owned by The Institute of Physics

Printed in Northern Ireland by The Universities Press (Belfast) Ltd.

Squire Tom Fut—Love's final gift,
remembrance

What we perceive as the present is the bright crest of an evergrowing past and what we call the future is a looming abstraction ever coming into concrete appearance. I love and revere the present. As to the past, my dealings with it are more complex, ranging as they do from delicious gropings to blind angry fumblings.

V. Nabokov

Preface

The solar system provides an exciting array of peculiar and individual atmospheric features. The colouring of Jupiter by complex organic molecules contrasts with the Earth's beautiful but transient rainbow. Consider, for instance, volcanic outgassing: enormous but apparently terminated activity on Mars contrasting with current activity on the Galilean satellite Io. A predominantly nitrogen atmosphere supporting clouds and possibly a methane 'ocean' on Titan makes a fascinating comparison to our own planet's atmosphere which is held in chemical disequilibrium by the action of the biosphere: the Earth has been described as 'a strange and beautiful anomaly in our solar system' (Lovelock 1979). This monograph is intended to assist and stimulate development of the history of planetary atmospheres and through this nurture better models of our own planetary environment.

Scientific study of planetary atmospheres and the atmospheres of major satellites draws upon both data currently being returned by spacecraft and upon theories of the origin of the solar system itself. It is a fascinating field which has contributed to our understanding of stellar evolution, climatic change and the origin of life. It is likely that the data presented here will be rapidly overtaken by newer sets of information. However, it is hoped that the theoretical models and methodology will serve as a useful framework within which to examine these new data sets.

The term planet will be used in a very wide sense in this work. It is important to define what the word will be intended to convey. It could be argued that any lump of material in orbit about the Sun is a 'planet' (or wanderer) in the solar system. However, very small objects will never possess more than transient atmospheres and are therefore omitted. A more precise definition of 'very small' is difficult since some planets (e.g. Mercury) do not retain significant atmospheres. Here a planet is taken to mean any body which has its shape dominated by the action of gravity (i.e. approximately spherical) and whose gravitational attraction is potentially significant compared with the scouring agencies for gases. This definition adds to the list of the nine planets, the following satellites: the Moon (Earth), Io, Europa, Ganymede and Callisto (Jupiter), Mimas, Enceladus, Tethys, Dione, Rhea, Titan and Iapetus (Saturn), Triton (Neptune) and Charon (Pluto).

Temperature is the fundamental characteristic of the planetary atmospheric environment. It determines the phase state of volatiles, the surface conditions and indeed the nature and extent of the atmosphere itself. This book is therefore primarily concerned with the determination of environmental temperature or planetary climate. The gross physical characteristics of the planet–atmosphere system determine this long-term climate and hence the chemical composition of the atmosphere and its evolutionary history. I shall discuss such physical characteristics as albedo, clouds, lapse rate dynamics and rotation which I believe must be well defined before, for instance, photochemical processes can be satisfactorily analysed.

Chapter 1 describes the origin of the Sun and solar system bodies. The mechanisms which control the atmospheric structure and characteristics at each stage in the evolutionary processes are described in chapter 2. Primary evolutionary histories of the planets and major satellites are discussed in chapter 3. These are then supplemented by a review of shorter-term 'climatological' changes in chapter 5. The fascinating evolutionary history of the Earth's atmosphere is described in chapter 4 with particular reference to the origin and evolution of life. Chapter 6 serves as a more personal epilogue in which some of the implications of the evolutionary histories described are examined.

Système International (SI) units are used in general but gas pressures are often expressed in millibars (1 mbar $= 10^2$ Pa) and time may be specified as '$\times 10^7$ years' or as '$\times 10^9$ years BP' where BP is the abbreviation for Before Present. Figures and tables are cross referenced throughout the text and variables defined as they are introduced. A full reference list is given at the end of the book.

I would like to acknowledge the help, stimulation and encouragement I have received over many years from Professor A J Meadows (Leicester University). He is an excellent scientist and a very good friend. I have also received counsel from Professors W H McCrea and J E Lovelock to whom I am very grateful. My husband, Brian, is a fantastic guy—words can't express my gratitude to and love for him and for his continuing encouragement and patience. My colleagues at Liverpool University have continued to support me both academically and personally for which I am most grateful. Substantial revisions were made to this text whilst I was a Visiting Research Associate at the NASA Goddard Institute for Space Studies, New York. I wish to thank all my colleagues there (especially Dr J E Hansen) and to acknowledge the financial and technical support of the National Research Council, USA. Finally, I am very grateful to friends and colleagues who read through draft versions of the manuscript and offered their comments; particularly Graham Cogley, Alan Schwartz, Jim Walker, Toby Owen, Jim Kasting, Rod Fujita and Dave (Bruce) Randall.

It is intended that this book will serve both as a text for taught courses and as a monograph for the research community. The diversity of the disciplines from which information is drawn may have resulted in a text which for any one scientist appears to contain juxtaposed concepts of variable levels of sophistication.

In view of such a rich diversity of problems to solve I am happy to admit considerable sympathy with the paradigm of the economy of scientific hypotheses (Occam's razor) i.e. if it is not necessary to invoke an exotic and individual explanation then it is preferable to avoid doing so. In this I may be seriously in error since, as a moment's consideration of the duck-billed platypus reveals, God does have a sense of humour!

A. Henderson-Sellers
New York 1981

Contents

1. The Origin of the Planets and the Evolution of the Sun

It is generally agreed that approximately 4.6×10^9 years ago the solar system formed from a rotating conglomeration of gas and dust. The formation and evolution of both the central star and the planetary components of our solar system are the subject of continuous and continuing debate. Cosmogony, the study of the origin of the solar system, has been said to demand 'the willing suspension of disbelief, no less than literary fiction' (Darius 1975). In the literature there exists a bewildering array of models of solar system evolution. Here we may accept the process as *fait accompli,* whilst acknowledging that the precise nature of planetary formation and stellar evolution will be of fundamental importance to all our succeeding discussions. Of particular importance are the chemical compositions of the terrestrial planets and the solar luminosity (see e.g. Lynden-Bell and Pringle 1974) and hence the black body temperature at each of the planet's orbital positions (see table 1.1). Williams (1975) has discussed the main features which must be explained by any acceptable model of solar system formation.

The Sun possesses practically all the mass of the solar system but only a fraction of the total angular momentum. The sum of the angular momenta of the planets, satellites and other smaller bodies is approximately 200 times greater than that of the slowly rotating central condensation. This suggests that the Sun lost much of its angular momentum during the formation process or during subsequent evolution. Any formation theory must be able to form at least nine planetary objects at great distances from the Sun. These

1

Table 1.1. Mass and composition of the planets (after Williams 1975).

Type	Planet	Orbital Distance (10^{11} m)	Mass (10^{24} kg)	Composition
Terrestrial	Venus	1.08	4.9	Fe, Si, O
	Earth	1.50	6.0	(Refractory type)
Major	Jupiter	7.78	1900	H, He
	Saturn	14.24	570	
Outer	Uranus	28.70	87	C, N, O
	Neptune	44.97	100	H, He
Others	Mercury	0.58	0.33	Fe, Si, O?
	Mars	2.28	0.64	
	Pluto	59.00	1.00	
Satellites	Ganymede	7.78	0.15	Fe, Si, O
	Triton	44.97	0.15	various ices
	Titan	14.27	0.14	
	Callisto	7.78	0.09	
	Io	7.78	0.08	
	Moon	1.50	0.07	

distances are related by the Titius–Bode law. The spacing of the planets is such that the mean distance of each from the Sun is approximately 75% greater than that of the next inner planet. The planets have widely ranging masses and compositions (table 1.1) but orbit the central condensation in a near planar configuration in a prograde sense. Data pertaining to the only other multi-bodied system yet observed (Black and Suffolk 1973), Barnard's star, are still controversial but may indicate that this alignment process is not necessarily an important attribute. The chemical composition of the planets is grouped in a manner associated with their masses and spatial positions.

1.1. Composition of the Planets

The planets were probably formed by the accumulation of interstellar grains and gases. In the case of the major and outer planets, the accreting mass was large enough to attract and retain gases. Table 1.2 (after Sagan 1975a) lists the main properties of all the planets in the solar system. (A minus sign

Table 1.2. Orbital and physical characteristics of the planets (after Sagan 1975a). Additional data for Pluto added from Trafton (1981).

	Mercury	Venus	Earth	Mars	Jupiter	Saturn	Uranus	Neptune	Pluto
Maximum distance from Sun (Millions of kilometres)	69.7	109	152.1	249.1	815.7	1507	3004	4537	7375
Minimum distance from Sun (millions of kilometres)	45.9	107.4	147.1	206.7	740.9	1347	2735	4456	4425
Mean distance from Sun (millions of kilometres)	57.9	108.2	149.6	227.9	778.3	1427	2869.6	4496.6	5900
Mean distance from Sun (astronomical units)	0.387	0.723	1	1.524	5.203	9.539	19.18	30.06	39.44
Period of revolution	88 days	224.7 days	365.26 days	687 days	11.86 years	29.46 years	84.01 years	164.8 years	247.7 years
Rotation period	59 days	−243 days (retrograde)	23 hours 56 min 4 s	24 hours 37 min 23 s	9 hours 50 min 30 s	10 hours 14 min	−11 hours (retrograde)	16 hours	6 days 9 hours
Orbital velocity (km s^{-1})	47.9	35	29.8	24.1	13.1	9.6	6.8	5.4	4.7
Inclination of axis	<28°	3°	23°27'	23°59'	3°05'	26°44'	82°5'	28°28'	17°10'
Inclination of orbit to ecliptic	7°	3.4°	0°	1.9°	1.3°	2.5°	0.8°	1.8°	17.2°
Eccentricity of orbit	0.206	0.007	0.017	0.093	0.048	0.056	0.047	0.009	0.25

Table 1.2. (*Continued*)

	Mercury	Venus	Earth	Mars	Jupiter	Saturn	Uranus	Neptune	Pluto
Equatorial diameter (km)	4880	12104	12756	6787	142800	120000	51800	49500	6000 (?)
Mass (Earth = 1)	0.055	0.815	1	0.108	317.9	95.2	14.6	17.2	0.0017
Volume (Earth = 1)	0.06	0.88	1	0.15	1.316	755	67	57	0.0133
Density (water = 1)	5.4	5.2	5.5	3.9	1.3	0.7	1.2	1.7	?
Oblateness	0	0	0.003	0.009	0.06	0.1	0.06	0.02	?
Atmosphere (main components)	None	Carbon dioxide	Nitrogen, oxygen	Carbon dioxide, argon (?)	Hydrogen, helium	Hydrogen, helium	Hydrogen, helium, methane	Hydrogen, helium, methane	Methane
Mean temperature at visible surface (°C) S = solid, C = clouds	350 (S) day −170 (S) night	−33 (C) 480 (S)	22 (S)	−23 (S)	−150 (C)	−180 (C)	−210 (C)	−220 (C)	−230 (?)
Atmospheric pressure at surface (mbar)	10^{-9}	90000	1000	6	?	?	?	?	?
Surface gravity (Earth = 1)	0.37	0.88	1	0.38	2.64	1.15	1.17	1.18	?
Mean apparent diameter of Sun as seen from planet	1°22′40″	44′15″	31′59″	21′	6′09″	3′22″	1′41″	1′04″	49″
Known satellites	0	0	1	2	13	10	5	2	1
Symbol	☿	♀	⊕	♂	♃	♄	♅	♆	♇

in front of the rotation period for Venus and Uranus indicates their retrograde rotation. Note also that since this table was compiled, the Pioneer and Voyager missions to the outer planets have added to the available data. For instance, the number of known satellites of Saturn is now 15.) Figure 1.1 is a schematic diagram showing the relative sizes of the bodies in the solar system. The considerable mass of the largest planet, Jupiter (Jupiter has a diameter approximately 11.2 times that of the Earth and its mass is $2\frac{1}{2}$ times the mass of all the other planets combined) is seen to be substantially less than the solar mass. It is interesting to note that, despite the fact that our solar system does not resemble a binary star configuration, the second most massive object is large enough

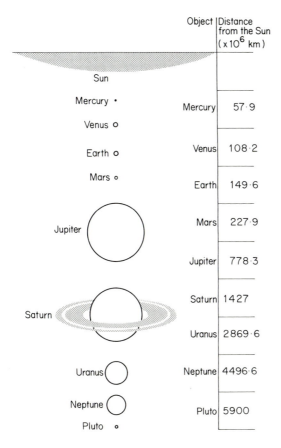

Object	Distance from the Sun ($\times 10^6$ km)
Mercury	57·9
Venus	108·2
Earth	149·6
Mars	227·9
Jupiter	778·3
Saturn	1427
Uranus	2869·6
Neptune	4496·6
Pluto	5900

Figure 1.1. Schematic diagram of the sizes of solar system bodies.

5

Table 1.3. Chemical reactions likely to take place in (i) equilibrium-condensation and (ii) inhomogeneous-accretion models of the solar system. The last two stages are unlikely to have occurred since temperatures have not fallen this low. The two models give rise to very different predictions for the chemical state of the planets (see also figure 1.2) (after Lewis 1974).

Degrees Kelvin	Equilibrium-condensation model	Inhomogeneous-accretion model
1600	1. Condensation of refractory oxides such as calcium oxide (CaO) and aluminium oxide (Al_2O_3) and also of titanium oxide and the rare-earth oxides	1. Condensation of refractory oxides such as calcium oxide (CaO) and aluminium oxide (Al_2O_3) and also of titanium oxide and the rare-earth oxides
1300	2. Condensation of metallic iron–nickel alloy	2. Condensation of metallic iron–nickel alloy
1200	3. Condensation of the mineral enstatite ($MgSiO_3$)	3. Condensation of the mineral enstatite ($MgSiO_3$)
1000	4. Reaction of sodium (Na) with aluminium oxide and silicates to make feldspar and related minerals, and the deposition of potassium and the other alkali metals	4. Condensation of sodium oxide (Na_2O) and the other alkali-metal oxides at about 800 K
680	5. Reaction of hydrogen sulphide gas (H_2S) with metallic iron to make the sulphide mineral troilite (FeS)	
1200–490	6. Progressive oxidation of the remaining metallic iron to ferrous oxide (FeO), which in turn reacts with enstatite to make olivine (Fe_2SiO_4 and Mg_2SiO_4)	
550	7. Combination of water vapour (H_2O) with calcium-bearing minerals to make tremolite	
425	8. Combination of water vapour with olivine to make serpentine	
175	9. Condensation of water ice	5. Condensation of water ice (H_2O)
150	10. Reaction of ammonia gas (NH_3) with water ice to make the solid hydrate $NH_3.H_2O$	6. Condensation of ammonium hydrosulphide (NH_4SH)
120	11. Partial reaction of methane gas (CH_4) with water ice to make the solid hydrate $CH_4.7H_2O$	7. Condensation of ammonia ice (NH_3)
65	12. Condensation of argon (Ar) and leftover methane gas into solid argon and methane	8. Condensation of solid argon (Ar) and methane
20	13. Condensation of neon (Ne) and hydrogen, leading to 75% complete condensation of solar materials	9. Condensation of neon (Ne) and hydrogen, leading to 75% complete condensation of solar materials
1	14. Condensation of helium (He) into liquid	10. Condensation of helium (He) into liquid

to generate significant internal energy through gravitational attraction (see chapter 3).

The mechanism of planetary formation is still the subject of debate (see the review by Williams 1975 and Ringwood 1979). The two extreme models of planetary formation generally discussed are (i) the equilibrium-condensation model and (ii) the inhomogeneous-accretion model. Table 1.3 (after Lewis 1974) lists the major reactions that would have occurred during the formation of the solar system under these two model histories. The two assumptions predict very different compositions for the terrestrial planets and the satellites resembling terrestrial planets. Figure 1.2 (after Lewis 1974) schematically compares the resulting composition of a number of the planets and satellites under these two hypotheses.

Recent data pertaining to the terrestrial planets seem to suggest that a formation sequence more closely resembling the inhomogeneous-accretion model probably took place (Benlow and Meadows 1977, Walker 1977, Owen 1978). Data pertaining to the Galilean satellites of Jupiter and particularly to Titan may prove to be extremely important in establishing the predominant formation mechanism of the planets. Wetherill (1980) has presented calculations which seem to overcome the few remaining obstacles to the acceptance of the accretion theory. Prior to his work it was acknowledged that unless the relative (collisional) velocities of the planetesimals could 'keep pace' with the increasing escape velocity, which increases as a function of radius, the accretional process would always be halted at some pre-planet stage. Figure 1.3 shows the results of Wetherill's calculations in which the average encounter velocity is the result of both collisional retardation and gravitational focusing. The escape velocity for the 830 km radius objects simulated is less than about $1.4\,\mathrm{kms^{-1}}$ in reasonable agreement with the simulated collisional speed.

In our investigation of climatic processes on the planets we shall be primarily concerned with atmospheric composition, tectonic processes, availability and surface interaction with volatiles. Table 1.4 (after Pollack and Yung 1980) lists the properties of atmospheres of solar system objects. The major

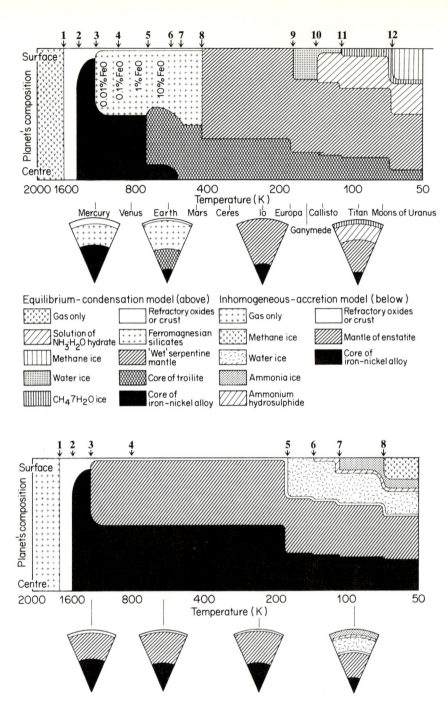

Figure 1.2. Lewis' (1974) results of different planetary formation models. The numbers relate to the major reactions in table 1.3. The pie diagrams represent the predicted core-to-mantle compositions of various of the planets and satellites (after Lewis 1974).

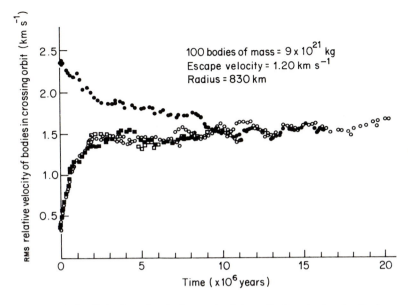

Figure 1.3. Calculated relative velocities of planetesimals during final stages of the planetary accretion process (from Wetherill 1980). The swarm's equilibrium velocity is found to be very close to the escape velocity and to be approximately independent of the initial velocity distribution.

division is between terrestrial and jovian† planetary atmospheres. The gaseous envelopes around the jovian planets are so close in composition to the solar nebula (Pollack and Yung 1980) that they may be considered primitive or original.

Current thinking about the early atmosphere of the Earth (and indeed all the terrestrial planets) supports the view long held by geologists (e.g. Abelson 1966) that the primitive atmosphere was an almost neutral or mildly reducing mixture of the volatiles CO_2, H_2O, N_2 and CO (Meadows 1972, Walker 1977, Owen *et al* 1979, Henderson-Sellers *et al* 1980). Noble gas data from the Viking mission to Mars reveals a very similar pattern of relative abundances as found on Earth. These ratios are also found in the 'planetary component' of meteoritic gases reinforcing the view that fractionation occurred prior to planetary formation. There is also a

† The traditional categorisation of the planets into a terrestrial or Earth-like group and a jovian or Jupiter-like group will be retained in this work despite the fact that data and theories discussed here underline the considerable differences within these two groups.

Table 1.4. Properties of the atmospheres of solar system objects (after Pollack and Yung (1980) in which the source references can be found). (a) Reading from left to right, the variables are the object's mean density (g cm^{-3}); acceleration of gravity (cm s^{-2}); surface pressure (bar); surface temperature (K), the numbers in parentheses are values of effective temperature; major gas species, the numbers in parentheses are volume mixing ratios; minor gas species, the numbers in parentheses are fractional abundance by number in units of ppm; aerosol species, the numbers in parentheses are typical values of the aerosols' optical depth in the visible. The numbers in this table were derived from the sources cited in Pollack and Yung (1980). (b) These mixing ratios refer to typical values at the surface. (c) These mixing ratios pertain to the region above the cloud tops. (d) The sulphuric acid aerosol resides in the lower stratosphere, while the sulphate, etc aerosols are found in the troposphere, especially in the bottom boundary layer. (e) The ice clouds are found preferentially above the winter polar regions. Dust particles are present over the entire globe. (f) These mixing ratios pertain to the stratosphere. NB Fink *et al* (1980) have reported the detection of a CH$_4$ atmosphere on Pluto (chapter 3).

Object	$\bar{\rho}^a$	g^a	P_s^a	T_s^a	Major gases^a	Minor gases^a	Aerosols^a
Mercury	5.43	3.95×10^2	$\sim 2 \times 10^{-15}$	440	He (~ 0.98), H (~ 0.02)[b]		
Venus	5.25	8.88×10^2	90	730 (~ 230)	CO$_2$ (0.96), N$_2$ (~ 0.035)	H$_2$O (20–5000), SO$_2$ (~ 150), Ar (20–200), Ne (4–20), CO (50)[c], HCl (0.4)[c], HF (0.01)[c]	Sulphuric acid (~ 35)
Earth	5.52	9.78×10^2	1	288 (~ 255)	N$_2$ (0.77), O$_2$ (0.21), H$_2$O (~ 0.01), Ar (0.0093)	CO$_2$ (315), Ne (18), He (5.2), Kr (1.1), Xe (0.087), CH$_4$ (1.5), H$_2$ (0.5), N$_2$O (0.3), CO (0.12), NH$_3$ (0.01), NO$_2$ (0.001), SO$_2$ (0.0002), H$_2$S (0.0002), O$_3$ (~ 0.4)	Water (~ 5) Sulphuric acid (~ 0.01–0.1)[d] Sulphate, sea salt Dust, organic (~ 0.1)[d]

Mars	3.96	3.73×10^2	0.007	218 (~212)	CO_2 (0.95), N_2 (0.027), Ar (0.016)	O_2 (1300), CO (700), H_2O (~300), Ne (2.5), Kr (0.3), Xe (0.08), O_3 (~0.1)	Water ice (~1)[e] Dust (~0.1–10)[e] CO_2 ice (?)[e]
Moon	3.34	1.62×10^2	$\sim 2 \times 10^{-14}$	274	Ne (~0.4), Ar (~0.4), He (~0.2)	—	—
Jupiter	1.34	2.32×10^3	$\gg 100^f$	(129)	H_2 (~0.89), He (~0.11)	HD (20), CH_4 (~2000), NH_3 (~200), H_2O (1?), C_2H_6 (~5)[f], CO (0.002), GeH_4 (0.0007), HCN (0.1), C_2H_2 (~0.02)[f], PH_3 (0.4)	Stratospheric 'smog' (~0.1), Ammonia ice (~1), Ammonium hydro-sulphide (~1), Water (~10)
Saturn	0.68	8.77×10^2	$\gg 100^f$	(97)	H_2 (~0.89), He (~0.11)	CH_4 (~3000), NH_3 (~200), C_2H_6 (~2)[f]	Same aerosol layers as for Jupiter
Uranus	1.55	9.46×10^2	$\gg 100^f$	(58)	H_2 (~0.89), He (~0.11)	CH_4	Same aerosol layers as for Jupiter, but thinner smog layer plus possibly methane ice
Neptune	2.23	1.37×10^3	$\gg 100^f$	(56)	H_2 (~0.89), He (~0.11)	CH_4	Same aerosol layers as for Jupiter, plus possibly methane ice
Titan	~1.4	$\sim 1.25 \times 10^2$	$2 \times 10^{-2} \rightarrow \sim 1$	~85	CH_4 (0.1–1)	C_2H_6 (~2)	Stratospheric 'smog' (~10)
Io	3.52	1.79×10^2	$\sim 1 \times 10^{-10}$	~110	SO_2 (~1)	—	—

Table 1.5. Volatile inventories of the Earth (present and sterile + no weathering states) and Venus (from Owen 1978 and updated from Hoffman *et al* 1980).

	Earth		Venus
Gas	Now	Total†	Now
N_2	78%	1.5%	3.4%
O_2	21	Trace	Trace
^{40}Ar	0.9	190 ppm	33 ppm
CO_2	0.03	98%	96.6%
Water	3 km	3 km	Trace
Pressure	1 atm	~70 atm	88 ± 3 atm

† No weathering, no life.

similarity between these volatiles and the gases observed in cometary tails although data likely to be furnished by the proposed fly-by mission (see Hughes 1980) of Temple-2 and Halley could be particularly useful here. The classic work of Rubey (1951) is now underlined by comparisons between the total volatile inventory of the Earth and the best estimate of the Venus inventory from Venera 9 and 10 (see table 1.5, after Owen 1978). However, any model of planetary formation must predict the mixing ratios for noble (or inert) gases in the terrestrial planetary atmospheres correctly. Recent observations of argon, krypton and xenon in the Venus atmosphere (Hoffman *et al* 1980, Donahue *et al* 1981) have cast serious doubts upon some solar system formation models (see chapter 3) and even suggest that volatile inventories may need to be reassessed.

The following propositions of Owen (1978) are the new orthodoxy (Henderson-Sellers *et al* 1980).

'1. The atmospheres of the inner planets are derived from volatile-rich veneers that coated these planets in the last stages of accretion.

 2. The primitive atmospheres of the inner planets were not highly reduced mixtures of CH_4, NH_3, and H_2, captured from the primordial solar nebula but only weakly reducing—containing CO_2, CO, N_2 and a small amount of H_2O, produced by degassing during and after accretion'.
     It is particularly interesting to note that all likely

mechanisms of atmospheric evolution result in an atmosphere dominated by CO_2 and H_2O and further that this composition does not conflict with Owen's final proposition that:

'3. The mean temperature of the early Earth was above 273 K, even though the solar luminosity was low and significant amounts of NH_3 were not present in the atmosphere'.

It should be noted that NH_3 cannot persist in any quantities in the early atmosphere even if it could be formed (Kuhn and Atreya 1979) and also that the lack of free iron in the upper mantle (e.g. Walker 1976, 1977) requires degassing of neutral or mildly reduced species CO_2, H_2O, CO, N_2, rather than reduced species CH_4, NH_3, H_2 for all models of planetary formation.

Despite this general consensus about the chemical nature of the primitive atmosphere, conflicts are common in the literature. There are two major reasons for suggesting the existence of highly reduced early atmospheres on the terrestrial planets. Many biologists and biochemists concerned with understanding the origin of life are still conducting investigations based upon the assumption of a CH_4–NH_3–H_2 atmosphere. However, as discussed in chapter 4, more recent laboratory work (e.g. Toupance *et al* 1978, Schwartz 1981) indicates that significant molecules can be built in more mildly reducing atmospheres. The nature of outgassed species is not easy to understand. It is possible that reduced gases currently vented to the atmosphere (Corliss *et al* 1981) are not juvenile. However Arculus and Delano (1980) have suggested that reduced gases may also have dominated earlier outgassing events although the data can be interpreted as indicating considerable mantle inhomogeneities. It is interesting to note that alternative theories regarding the oxidation state of the early atmosphere originating both from geological/geophysical data (e.g. Arculus and Delano 1980) and from biochemical data and experiments (e.g. Corliss *et al* 1981) seem to have an early hydrosphere as a common element. As described in chapter 4 rapid removal of reduced species, particularly NH_3 and CH_4, is inevitable in a water dominated global environment. It is uncertain how or at what rate the Earth's atmosphere evolved but the resultant chemical state must have been neutral or mildly reducing.

The rate of atmospheric build-up is discussed in chapter 2. Chemical and geophysical processes clearly warrant detailed consideration since, for instance, the removal of gases (e.g. CO_2) from the atmosphere not only affects the total atmospheric mass but also decreases the greenhouse effect[†]. Chemical reactions (especially photochemical ones) may considerably perturb levels of trace gases in the primitive atmosphere and, not withstanding the foregoing argument, small amounts of NH_3 and CH_4 may have been transiently present as a result of meteoritic bombardment (see chapter 4).

The evidence for a 'secondary' or derived atmosphere around each of the terrestrial planets is based upon data pertaining to noble gas fractionation relative to solar abundances (see chapter 3). Various mechanisms (e.g. the T-Tauri stage of the Sun) have been proposed for stripping the inner planets of any remaining hydrogen, helium, etc (Cameron and Pollack 1976). Here it is of little importance to attempt to establish the reason for the lack or loss of a 'primary' or 'primitive' atmosphere for these planets. The nature and rate of the build-up of the secondary atmosphere will be seen to be critical in the discussion of climatic and surface (i.e. bombardment and tectonic processes; see e.g. Mutch 1979) processes in chapters 3 and 5. It is therefore important to consider the differences between the inner planets and how these physical parameters might relate to core formation, mantle circulation and volcanic/tectonic degassing.

1.2. Solar Evolution

The temperatures of all the terrestrial planets are directly dependent upon the radiant energy available to them from the Sun. This energy is modified both by the planet–Sun distance and by the reflectivity of the planet. One of the most important factors affecting atmospheric evolution is the perturbing effect of any modifications in solar luminosity.

---

† Planetary surface temperatures are often higher than the effective or equilibrium temperature. This enhancement is the result of absorption of thermal radiation emitted from the surface and is termed the greenhouse effect (see chapter 2).

By the late 1950s, the models of stellar evolution and, particularly, of the evolution of the Sun were well established. Age estimates made for the Sun gave values of approximately 5.0×10^9 years, which were in good agreement with independent estimates for the age of the solar system. The Sun was known to be a typical G2 star, and its probable evolutionary track along the Main Sequence had been calculated. These calculations permitted estimates to be made of the likely luminosity change for the Sun over its lifetime. The estimates differed slightly but gave, in general, consistent results. For instance, Schwarzschild *et al* (1957) gave

$$L_{present}/L_{initial} \simeq 1.6 \qquad (1.1)$$

whilst, from a similar calculation, Haselgrove and Hoyle (1959) suggest that this value was about 1.46. The second of these ratios was an improvement, because the model calculation took into account small effects neglected in the earlier work. This ratio of present luminosity to initial luminosity gives a percentage decrease (from the present value) in the luminosity of the Sun, over its lifetime, of approximately 30%. The values computed for the luminosity divided by the present luminosity are illustrated in figure 1.4 as a function of

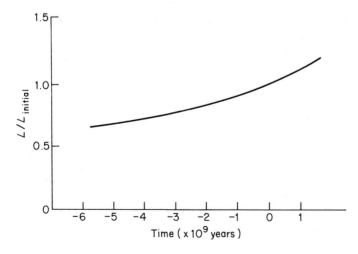

Figure 1.4. Variation of the ratio of solar luminosity to present-day luminosity throughout the lifetime of the solar system (based on Haselgrove and Hoyle 1959).

time. Thus the percentage increase in the solar luminosity over the lifetime of the planets (say 4.5–4.6×10^9 years) is approximately 43%.

It has been suggested that such a large variation in the solar luminosity could have been responsible for detectable terrestrial temperature changes (see, for instance, Schwarzschild 1958). The magnitude of this luminosity change led Sagan and Mullen (1972) to compute temperatures for the early Earth that were unacceptably low, and, consequently, to postulate the existence of additional absorbing gases in the early atmosphere. This calculation omitted several very important considerations (see chapters 3 and 4).

Since the development, and general acceptance, of these models of solar evolution, observational data have created a number of problems. Since 1968, an experiment has been under way to try to count solar neutrinos. Neutrinos were calculated to comprise about 3% of the total energy released by the Sun. The advantage of direct capture of these particles, which are produced during weak decays of nuclei in the central core of the Sun, was that for the first time data would be available on activity deep inside the Sun. The experimental problems of detection are very great. The technique adopted by Davis (see Wick 1971, Bahcall and Davis 1976) is to capture the neutrinos by interaction with chlorine-37 isotopes and count the resulting argon-37 atoms. The earliest results published by Davis and his co-workers indicated a rate of detection of neutrinos considerably lower than that predicted by the conventional solar models (Wick 1971). Improvements in the experimental techniques seem to have led to much higher neutrino counts (Bahcall and Davis 1976), though the new counts still fall short of the predicted values, and the average over the whole experimental period remains below expectation (Davis 1979). Figure 1.5 illustrates Davis' results and indicates the predicted level. Recently it has been suggested that the neutrino may possess mass and can 'oscillate' between any of three types (Waldrop 1981). Since the Davis experiment was designed to capture only one of these types, the electron neutrino, it is possible that his results are reconcilable with the predictions of modern stellar models.

Newman and Rood (1977) have reviewed the predictions

of solar luminosity of both standard and more recent stellar models. They emphasise that increasing luminosity is a fundamental feature of almost all theoretical models. The rate of luminosity change with time reduces to the simple form

$$\frac{1}{L}\frac{dL}{dt} \simeq \frac{7.5}{\mu}\frac{d\mu}{dt} \qquad (1.2)$$

where μ is the mean molecular weight of the star. (Note should be made here that equation (1.2) is not valid if the gravitational constant changes with time (e.g. Canuto and Hsieh 1980).) Newman and Rood (1977) list values for $L^{-1}\,dL/dt$ derived from a large selection of stellar models. These range from 0.013 to $0.72/10^9$ years with the standard stellar models suggesting approximately $0.05/10^9$ years. Thus Newman and Rood (1977) advocate

$$L_{present}/L_{4.5} \simeq 1.25. \qquad (1.3)$$

It appears that theoretical and observational work in the last few years have re-established the credibility of solar evolution models; e.g. Christensen-Dalsgaard and Gough (1976, 1981) report observations by themselves and Grec et al (1980) which seem to agree well with theoretical results.

Temperature curves for the planets may therefore be computed, assuming that the solar luminosity has increased by between 25 and 45 per cent from its initial value when the

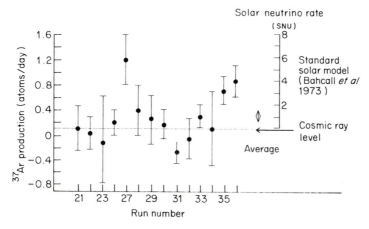

Figure 1.5. Solar neutrino counts (data from Bahcall and Davis 1976). The average count level is below the predicted values.

planets condensed over the lifetime of the solar system (chapter 3).

Stellar evolution involves changes in the nature of the energy emitted from the star as well as the total flux. The variation of the wavelength, w_λ, of peak emission is obtained directly from Wien's displacement law which gives the position of the peak of the Planck curve for the effective stellar temperature $T(K)$ as:

$$w_\lambda = 4.0947 \times 10^{-12} \, T^5 \quad (\text{W m}^{-2} \mu\text{m}^{-1} \text{sr}^{-1}). \qquad (1.4)$$

Thus Planck emission curves for various stages of stellar evolution may be produced following the calculation of T. (NB T varies much more slowly than luminosity, L.) Furthermore, currently the emission in the ultraviolet part of the spectrum does not follow the Planck curve for a star of the Sun's effective temperature ($\simeq 5700$ K). During the earliest history of the solar evolution this UV flux could have been greatly enhanced (e.g. Imhoff and Giampapa (1980) note very much increased UV fluxes for T-Tauri stars). It is possible that a simple linear scaling of the ultraviolet radiation with the effective temperature of the Sun is unwise, at least for the very earliest stages of solar evolution. Since the UV flux is often cited as an important factor in both atmospheric photochemistry and in the origin and evolution of life this difficulty in estimating the likely fluxes may become important.

1.3. Format of Study

Chapter 1 has set the scene within which the origin and evolution of the atmospheres of the planets of the solar system will be examined. The important features common to all the atmospheres are discussed in detail in chapter 2. Major interaction between atmospheric constituents and a planetary surface is of fundamental importance in the evolutionary process. The planets upon which such interactions occur are treated separately from the mainly gaseous planets in the discussions in chapter 3. A fascinating comparison between Titan and the Earth is drawn. Many of the planetary atmospheres seem to have suffered very early upheavals but the subsequent evolutionary paths can be traced. Chapter 4 deals with the Earth, particularly with the problem of the origin and evolution of life and their effects upon the atmosphere and climate. Comparative planetary climatology (chapter 5) is useful both in our attempts to study long-term evolutionary processes and as an aid to interpreting the climatic history of the Earth. Chapter 6 presents a more personal view of the uniqueness of our planet in the form of an epilogue.

The investigation of planetary atmospheres presented here seems to suggest that planetary environments which are controlled by latent heat exchanges in a substantial atmosphere exhibit the greatest stabilities. Common processes and common features can be examined in a way that leads to a more complete understanding of the solar system as a whole and of the Earth itself.

2. Evolution of Atmospheres: The Mechanisms of Long-Term Change

The principal characteristics of any atmosphere are its mass, chemical composition, motion relative to the parent body and vertical structure. It is very important to differentiate between rapid and/or transient (i.e. reversible) atmospheric changes and long-term planetary evolution. This text is an attempt to describe the evolutionary climatology of planetary atmospheres. The fundamental characteristic of the atmospheric climate is temperature. This can either be the cloud-top temperature (e.g. Jupiter) or the surface temperature (e.g. the Earth). There is a fundamental link between these two temperatures which is critical for the most interesting group of terrestrial-type planets (§3.4). It is the relationship between the evolutionary state (typified by temperature) and the physical characteristics of the planet with which this work is primarily concerned.

A number of excellent texts have reviewed chemical and photochemical reactions, upper-atmosphere processes and escape, geological and biological data. Among the most recent are McElroy (1976), Walker (1977), Holland (1978), Levine and Schryer (1978) and Dastoor *et al* (1981) upon which I have drawn extensively in constructing chapters 3–6. Detailed analysis of processes not generally considered in an evolutionary time frame (e.g. planetary rotation rate and

20

dynamics; cloud type, formation and extent) is given here. Other topics of equal importance are only briefly reviewed (e.g. §2.2.2).

It is likely that changes in any atmospheric characteristic will also result in changes in other characteristics. For instance, addition to the atmosphere of new gases possessing infrared absorption bands will increase the greenhouse effect and by changing the surface temperature result in a new vertical temperature structure, more clouds and possibly, through evaporation or sublimation of volatiles from the surface, an increased overall atmospheric mass. The interactions between the atmosphere and the planetary surface are clearly very important.

The other primary location of change is the exosphere†. The Earth's exosphere has a base located at a height of between 400 and 500 km. Practically all additions and removals from the atmosphere take place either at the surface or in the exosphere. Chemical processes occur throughout the atmosphere and at the surface whilst physical processes, such as condensation of volatiles, are almost entirely restricted to the troposphere or mixed layer and the surface.

The importance of the surface as a site for chemical and physical changes is clear. The evolutionary histories for the major (or jovian) planets will concern us less than those of the terrestrial planets for two reasons. The first is that either a planetary surface does not exist or that it is far removed from the atmosphere. Secondly the outer planets have such large masses, and hence gravitational attraction (see table 1.1), that escape from the exosphere is very much less likely than from smaller bodies such as satellites and the terrestrial planets. Even the lightest element, hydrogen, is retained. The lack of a surface and a large enough overall mass that makes escape minimal will place the parent body in a category of little evolution whereas any other system will be likely to suffer considerable evolutionary changes. An interesting example of the importance of a surface is the outermost planet. Pluto is now known to possess a tenuous CH_4 atmosphere

---

† The exobase is the height in an atmosphere above which e^{-1} (approximately 36.8%) of the upward moving gaseous particles will not experience a collision.

(Fink *et al* 1980). The presence of this atmosphere appears to be related to the vapour pressure of methane over its solid phase at temperatures in the same region as the subsolar temperature ($\simeq 60$ K), see chapter 3.

In this chapter all the major chemical and physical processes operating in atmospheres are described briefly but particular attention is paid to the mechanisms which will control any evolutionary changes. The reason for considering these processes is that all the important evolutionary steps are the result of the interplay between external and internal forces and the planetary environment. It is for this reason that the processes and the evolutionary steps are considered together throughout this work. This approach is more complex than the traditional listing of sources and sinks of atmospheric volatiles followed by brief consideration of possible planetary histories but, in the light of the most recent data, seems to be more worthwhile.

Planetary and satellite atmospheres have certain fundamental characteristics in common: they are gravitationally bound around an approximately spherical parent. The parent rotates and is illuminated by the Sun around which it revolves. The constituents of the atmosphere are in their gaseous phase at approximately the mean surface temperature. They can, and do, interact both chemically and physically with one another and the surface. Even a list as basic as this of the characteristics of atmospheres illustrates the importance of the *differences* between the planets and satellites and hence their atmospheres. For instance the very existence of a surface is of critical importance to both long- and short-term atmospheric changes (see chapters 3 and 5) particularly as this is often the site of phase changes of important volatiles. For the smaller planets and moons the gases are also affected by the gravitational attraction of the parent planets but often not bound by it. This is also true of the upper atmospheres of all the planets. Total interactions between the parent planet and other astronomical bodies may be important. Internal feedback effects between gases and the surface are capable of modifying the atmosphere.

Table 1.4 lists the properties of the atmospheres of the planets in the solar system together with a number of fun-

22

damental planetary properties. The major gaseous constituents reflect the sources of volatiles as described in chapter 1; the terrestrial planets having CO_2, H_2O and N_2 dominated atmospheres whilst the massive jovian planets have retained their original envelopes of light gases such as H_2 and He. The current nature and amount of atmosphere is a function of the whole evolutionary process which will be considered in chapter 3. Table 2.1(a) (from Pollack and Yung 1980) lists the sources of volatiles for the atmospheres.

The noble or rare gases (He, Ne, Ar, Kr, Xe) are particularly interesting elements of the planetary atmospheres. This is because they are chemically inert. Thus the assumption can be made that, after excluding losses by escape, the inventory of noble gases observed today is the direct result of either protoplanetary formation or crustal release although there is the possibility of gain of noble gases from space (see §2.1.5). For the major planets even the lightest noble gas, He, is a valuable datum point but the loss rate of He from the terrestrial planets is fast enough that its primary use is as a measure of continuing outgassing. Comparison of the ratios of isotopes of a selected species can provide direct data on the degassing extent and rates when one isotope is a radiogenic daughter. For instance, ^{40}Ar results from the radioactive decay of ^{40}K which is bound into the planet. Thus the ratio of $^{36}Ar:^{40}Ar$ is a useful guide to the extent of degassing. Detailed intercomparison of noble gas ratios between the planets and the Sun can provide an estimate of the total volatile inventories. Table 2.1(b) (from Walker 1977) lists the relative abundances of some elements on the Earth compared with the whole solar system.

The mechanisms which control the atmosphere's physical and chemical state at any point in its evolution are clearly important since it is this state which determines the likelihood and extent of escape, addition and interaction at the surface. The structure and composition of atmospheres is reviewed below. Dynamics are much less important on evolutionary timescales than the average static conditions. However, the relative magnitude of the radiative and dynamic energy transport mechanisms in the planet–atmosphere system and indeed the way that the planet 'looks' (e.g. the

Table 2.1. (*a*) Potential sources of key volatiles for planetary atmospheres (after Pollack and Yung 1980). For sources and method of estimating relative importance, see Pollack and Yung (1980). Note that the solar wind remains the major source of volatiles for Mercury and the Moon. (*b*) Abundances of selected elements in the Earth compared with the whole solar system abundances (after Walker 1977, from Mason 1958).

(*a*)

Source	Elemental composition	Time period of greatest potency
Primordial solar nebula	H > He > O > C > N > Ne > Mg > Si > Fe > S > Ar	4.6×10^9
Solar wind	H > He > O > C > N > Ne > Mg > Si > Fe > S > Ar	$\sim 4.6 \times 10^9$ decreasing to date
Colliding bodies	H_2O > S > C > N	$\sim 4.6{-}4.0 \times 10^9$
Interior	H_2O > C > Cl > N > S	$\sim 4.6{-}4.0 \times 10^9$

(*b*)

	Atomic number	Whole Earth (atoms/10 000 atoms Si) (*a*)	Solar System (atoms/10 000 atoms Si) (*b*)	Deficiency factor (log (*b*/*a*))
H	1	84	3.5×10^8	6.6
He	2	3.5×10^{-7}	3.5×10^7	14
C	6	71	80 000	4.0
N	7	0.21	160 000	5.9
O	8	35 000	220 000	0.8
F	9	2.7	90	1.5
Ne	10	1.2×10^{-6}	50 000	10.6
Na	11	460	462	0
Mg	12	8 900	8 870	0
Al	13	940	882	0
Si	14	10 000	10 000	0
P	15	100	130	0.1
S	16	1 000	3 500	0.5
Cl	17	32	170	0.7
Ar	18	5.9×10^{-4}	1 200	6.3
Kr	36	6×10^{-8}	0.87	7.2
Xe	54	5×10^{-9}	0.015	6.5

bands and zones and the Great Red Spot on Jupiter, the mid-latitude depression systems on Earth) may change significantly over evolutionary time periods. The fascinating recognition that all the terrestrial planets seem to have established regimes within which the surface environment remains close to the sublimation/evaporation condition of a primary atmospheric volatile can only be fully understood by reference to dynamic as well as static environmental controls. The regimes of the Earth, Mars and Titan are controlled by phase changes of the volatiles H_2O, CO_2 and CH_4, respectively. On Venus chemical equilibrium between CO_2 and the surface has been established and SO_4^{2-} species change physical and chemical states readily.

2.1. Mechanisms and Rate of Atmospheric Build-up

Tectonic activity and impacts from outside the planet are the major processes by which an atmosphere is formed and modified. The two processes are, to first order, independent of one another but very large impacts may produce tectonic activity. The other agents of tectonism are convection in the mantle and gravitational interaction between the planet and a parent body. Tidal forces between the Sun and the planets are too weak to cause crustal upheavals so that latter form is important only on satellites. The nature of these processes will depend upon the extent of the atmosphere. For instance, the likelihood of meteoritic vaporisation diminishes as the atmospheric mass increases. To a lesser extent the effect of tectonic activity is also dependent upon the atmospheric mass but in this case it is the state of the planetary interior which is crucial.

The amount and rate of bombardment of the newly accreted planets and satellites during the first 0.5×10^9 years of the solar system (see chapter 1) may be important not only for establishing the initial extent and chemical composition of the atmospheres but also for determining the nature of surface features produced. The Voyager spacecraft missions to Jupiter and Saturn have sent back data which appear to confirm the hypothesis that bombardment was extensive but declining throughout the time period $4.5–4.0 \times 10^9$ years BP.

This intensive bombardment is significant for the surface and mantle states (see figures 2.1(*a*) and (*b*)). Furthermore, Windley (1977, 1980) has suggested that it coincided with the period of important radioactive heating within the planetary body. Figure 2.1(*b*) is a schematic representation of the probable curves of impact and radioactivity. Windley (1980) states that for the Earth the period between 3.8 and

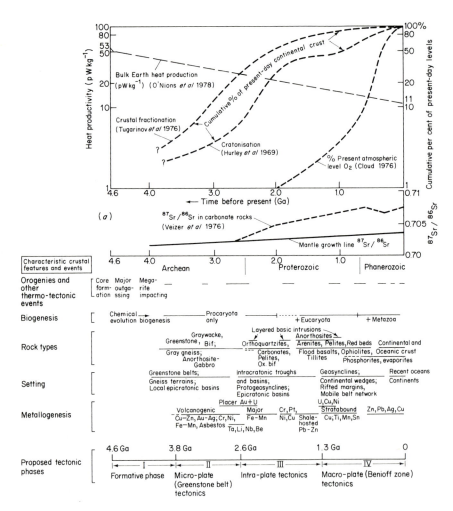

Figure 2.1. (*a*) The development of the Earth's crust as a function of time ($\times 10^9$ years); crustal controls, trends and major periods are shown (cf figure 2.3) (after Goodwin 1981), see source for references. (*b*) Schematic diagram of the relative importance of energy fluxes to planetary surfaces. It is probable that there was a late peak in bombardment. The two sources are both high around 4.1×10^9 years BP.

26

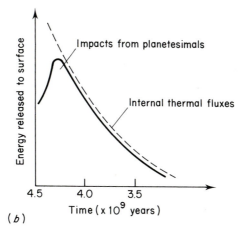

(b)

Figure 2.1. (*Continued*)

3.0×10^9 years BP was the time of rapid crustal accumulation. This growth, which, by analogy with the present-day situation, is likely to have occurred by the creation of new continental material at mid-oceanic ridges, must have drawn its energy from the breakdown of radiogenic isotopes (^{40}K, ^{238}U, ^{235}U and ^{232}Th) trapped in the planetary core. It is believed that currently over 70% of the thermal energy released by such decay is directly used in crustal formation and plate movement. Energy released by radiogenic decay will decrease with time but observational data suggest that there was a significant increase in the bombardment of the planets and satellites around 4.1×10^9 years BP (figure 2.1(*b*)). The combination of these two energy sources at about this time probably caused considerable tectonic activity on the Earth (figure 2.1). It is possible that these two processes combined to give rise to approximately three times as much tectonic energy as at the present time. There are likely to have been more plates in the lithosphere and their movement may have been somewhat faster. It is less clear what effect this peak in energy may have had for the other terrestrial planets (see §3.4). Surface bombardment will produce observable features if there is no atmosphere, or if it is very tenuous or if the impact is extremely large. These features are readily observable in a panorama of the solar system (Guest *et al* 1979) and include the impact crater itself

27

with its attendant terraces, dunes and secondary craters. There is also the possibility of regolith shock causing volcanic effusions. Any such impact features will be rapidly removed in an environment like that of the Earth through the combined effects of tectonics, erosion and atmospheric weathering. A similar but much weaker effect is seen on Mars where 'fluvial' features appear to have occurred at the sites of earlier impacts. Presumably if the climatic epochs which led to this flow of surface water had been sustained the impact features would ultimately have been removed.

It is therefore very interesting to note the suggestion of Guest (1980) that the ring of mountains, in Aphrodite Terra, on the planet Venus may be the result of major impacts which occurred during the period of intensive bombardment around 4.0×10^9 years BP (see also Pettengill *et al* 1980). If this topographic feature is shown to be such an impact feature than it seems reasonable to deduce that neither global-scale tectonism nor large-scale (and especially hydrospheric) atmospheric weathering have ever occurred on Venus. The latter deduction will be extremely important in any history of the atmospheric evolution of Venus (see §3.4). Furthermore, the conditions in the forming solar system and on the juvenile surface are critically important in determining the final chemical and physical (i.e. phase) state of any volatiles acquired.

2.1.1. *Impact Degassing*

It has been suggested that the atmospheres appeared over a brief time span, either as part of the planetary formation process, or not too long after its occurrence. Although rapid atmospheric production on the terrestrial planets might, theoretically, be triggered in a number of ways, the most physically feasible mechanism is by impact vaporisation during the final stages of planetary accumulation. There is quite strong evidence for an early rapid degassing event (Ozima 1975). (This does not rule out later additional degassing, so slow degassing models need not necessarily be abandoned, see §2.1.2.) For example, Benlow and Meadows (1977) argue as follows:

The original solar nebula, being of moderate density,

28

should have sorted protoplanetary material into approximately circular orbits. Relative velocities in the nebula would therefore have been small, and material accreted during the initial stages of planetary formation would not have vaporised extensively on impact. This situation could change once the gravitational attraction of the forming planet became significant.

Velocities of a few kilometres per second seem to be necessary for vaporisation (Ahrens and O'Keefe 1972, see also, Wetherill 1980). Benlow and Meadows (1977) define a mass band around 5×10^{23} kg and expect that, if impact vaporisation played a major role in atmosphere formation, planets with masses above this band would have extensive atmospheres, whilst those below the band would not. The observational data, they suggest, fit this picture quite well.

The appearance of the terrestrial atmospheres must be connected with the final stages of planetary accretion, so it is anticipated that any material still remaining from planetary formation reflects the composition of this final fraction i.e. similar to C1 carbonaceous chondrites and certain comets, implying that the commonest volatiles generated by impact should be CO_2 and H_2O (table 2.1(a)).

The results of Benlow and Meadows (1977) for the total amount of volatiles degassed on Earth cluster around a value of 10^{22} kg and prove to be rather insensitive to variations in the input data. They compare this value with an estimate for the current mass of terrestrial volatiles (atmosphere + hydrosphere + sediments) of 4×10^{21} kg (Ronov and Yaroshevsky 1967) and suggest that an impact origin for the atmosphere is feasible.

It is important to examine how such rapid formation would have affected the surface temperature. Impacts must presumably enhance the surface temperature at least locally, but, as we have seen, the number of surface impacts decreases with time and the energy input by the infalling bodies is increasingly transferred to the upper atmosphere, where it can be dissipated readily†. It is interesting to speculate on the

† NB Walker (1977) has shown that accretional heating can only be significant for comparatively short times ($<10^6$ years) through potential energy considerations alone; but impacting material possessing significant kinetic energy could be an additional heat source.

possible climatic effects such a vertical variation in net heating may have had. Heating by impact of meteoritic-type material varies inversely with the amount of atmosphere present, and radiation of the surface heat to space occurs quite rapidly at all stages. The main problem in determining the early surface temperature consequently concerns the nature of the atmosphere itself and more especially the possibility of a runaway greenhouse effect occurring. This may be critical for the Earth under this extremely rapid model for atmospheric evolution. Although the black body temperature in the neighbourhood of the early Earth was below the freezing point of water, the surface temperature of the planet would have increased as the atmosphere grew. The magnitude of this temperature increase depends upon the amount of carbon dioxide left in the atmosphere if all the water condenses into oceans. This amount in turn depends on the chemical state of the primitive oceans. On the early Earth, it is probable that the formation of carbonates and bicarbonates would have been with alkali metals. Rapid precipitation would still have resulted, leading to a lowering of the residual atmospheric CO_2. The final equilibrium figure calculated by Benlow and Meadows (1977) is close to or less than 1000 mbar CO_2 causing higher temperatures but not, they suggest, a runaway greenhouse effect. This result is supported by the work of Owen et al (1979) who suggest that an enhanced CO_2 greenhouse is sufficient to overcome the lowered solar luminosity.

Over longer geological time periods the contribution by impacts to the total volatile inventories is likely to be much less significant than during the final accretion phase. Pollack and Yung (1980) calculate that assuming an impact history for the Earth similar to that for the Moon a total contribution of only 2×10^{18} kg of water can be expected from later meteoritic bombardment. This is negligible compared with the current oceanic mass ($\approx 1.5 \times 10^{21}$ kg H_2O).

2.1.2. Surface (Slower) Degassing

There are likely to be three phases in the release of volatiles to the atmosphere: (a) during accretion; (b) at or following accretion due to the heat flow generated by planetary differentiation and (*c*) as a result of local (but possibly extensive) volcanism. The general consensus is that most of the major outgassing occurred during the first 1×10^9 years after planetary formation (e.g. Walker 1977). It is important to note that the degree, and also possibly the timing, of this outgassing may be a function of the planetary mass. Size dependence results from the smaller levels of both the energy gained during accretion and the lower flux from the decay of radiogenic nucleii (e.g. Jakosky and Ahrens 1979).

Over time periods of 10^9 years the effect of atmospheric changes on the surface temperature is difficult to calculate. The assumption of planetary radiation balance defines the effective planetary temperature, T_e such that

$$S(1-A) = f\sigma T_e^4 \tag{2.1}$$

where S is the solar flux, integrated over all wavelengths, at the average distance of the planet from the Sun (i.e. the solar constant in the case of the Earth); A is the Russell–Bond spherical albedo for the planet; f (the 'flux factor') represents the ratio of the area of the planet emitting radiation to the area intercepting the flux of solar radiation. The surface temperature is related to the effective temperature via the greenhouse increment: calculating the connection between the two requires a knowledge of the atmospheric mass and chemical composition. Henderson-Sellers and Meadows (1979b) have developed a numerical scheme which relates these variables and thus can be used to produce a series of atmospheric models.

In principle the surface temperature can be calculated for any given atmospheric mass. Over time periods of about 10^9 years other factors cannot be assumed constant. The solar flux has almost certainly increased over the lifetime of the solar system (see §1.2). Changes in the albedo are still more problematical. For slow degassing, the planet is initially almost without an atmosphere and the albedo will be low. As the amount of atmosphere and the percentage cloud cover builds up, so the albedo increases (see §2.5). Hence the albedo must be expected to be related in some way to the stage of degassing that has been reached. Finally, the flux

31

factor is normally considered to be constant, with a value either equal to two, for a slowly rotating planet with little atmosphere, or four, for a rapidly rotating planet with a thicker atmosphere. However, if an atmosphere builds up slowly, the value to be assigned to the flux factor requires more careful consideration, see Henderson-Sellers and Meadows (1976) for a fuller discussion. This is especially true for the Earth, since the gravitational interaction of the Earth and the Moon has probably changed the terrestrial rotation rate considerably since the Earth first formed.

The flux factor for the Earth can vary in a quite complex way during its early history, if the changing rotation rate is coupled with a slow build-up of the atmosphere. The assumption of slow degassing produces the variation of surface temperature with time for the Earth shown in figure 2.2. That an early peak in temperature is produced by both rapid and slow, degassing is interesting, since some geochemical

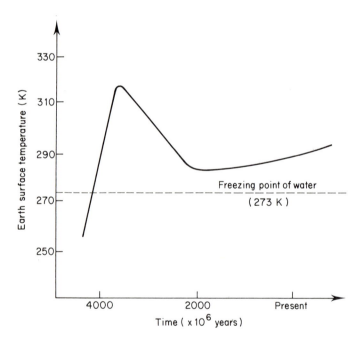

Figure 2.2. Calculated curve of the Earth's surface temperature (after Henderson-Sellers *et al* 1980). The early peak in T_s is a result of slow degassing and the changing rotation rate of the Earth (see text).

32

measurements suggest that global temperatures may have been higher during the Precambrian era (Knauth and Epstein 1976).

It is also important to note that these predictions of a high early value of T_s derive from different effects. In the rapid degassing case T_s is increased by large quantities of CO_2 while in the slower degassing case the peak in T_s is the result of the changing physical characteristics of the planet (see also §3.2).

The evolutionary histories of the atmospheres around all the terrestrial planets almost certainly follow paths between the two extreme degassing models. Since we know that practically all the volatiles have been derived from the planetary material itself (there may have been enhancement in the more tenuous regimes through interaction with the solar wind) different evolutionary histories may be caused either by the difference in the planetary compositions or by the predominance of different mechanisms of control. These controlling factors include:

(a) the amount, type and rate of degassing activity;
(b) the mode of planetary formation itself (chapter 1);
(c) chemical reactions and loss rate of gases to space; and
(d) internal feedback mechanisms altering such things as total atmospheric mass, albedo, flux factor, etc.

It is generally believed (see chapter 1) that the terrestrial planets accreted rapidly. Thus internal planetary temperatures must have been high and it now seems reasonable to assume that the formation of a metallic iron core is a fundamental part of the planetary formation process. The effect of alternative histories on the chemical state of gases evolved is described in §2.2. Walker (1978c) suggests that as the central regions of the planets cooled all would have progressed through similar tectonic activity states (figure 2.3). The rate of tectonic evolution (figure 2.4) seems to be a function of the planetary mass (and also of total planetary composition) and will be fundamental in determining the amount and state of degassed material (e.g. Toksoz *et al* 1978). Weathering of surface rocks of course removes atmospheric gases but a tectonically young and active planet will also recycle the resulting sedimentary material very much more rapidly than a

33

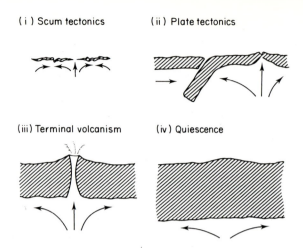

Figure 2.3. Possible sequence of tectonic activity on a terrestrial-type planet (after Walker 1978c).

planet in the later stages of terminal vulcanism or quiescence. It is possible that, whilst the smaller masses of Mercury, the Moon and indeed Mars have led to comparatively early and now almost completed tectonic, and hence atmospheric histories, the major cause of the contrasting evolutionary histories of Venus and the Earth (with planetary masses almost

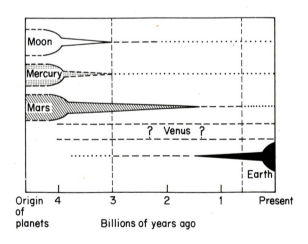

Figure 2.4. Ages of planetary surface units as a function of time ($\times 10^9$ years). The plotted area is the estimate of the present surface of that age. The early tectonic activity on the smaller planets contrast strongly with the Earth (after Head and Soloman 1981).

34

identical ($\sim 5 \times 10^{24}$ kg)) is a result of the very high surface temperatures upon Venus (see chapter 3).

The limited evolution of the atmospheres of the jovian planets will be even more strongly controlled by factors (*b*) and (*c*) above since degassing has not occurred and interaction with lower levels and with the external environment are of less significance for these giant planets. The planetary evolution with time may cause changes in the composition and even the mean temperature of these atmospheres (chapter 3).

2.1.3. Dynamics

Atmospheric motions are the result of radiation forcing causing motion in a rotating and gravitationally controlled system. Radiation imbalances can lead to vertical motion (see §2.1.4) but they are also responsible for horizontal movement. Radiative forcing arises, for example, from the surface temperature variation between the day and night sides of any planet with a fairly tenuous atmosphere (e.g. Mars) and from the annually averaged imbalance of net radiation e.g. the Earth (see figure 2.5).

Diurnal temperature changes on the Earth are of little importance for planetary-scale motion since the thermal relaxation time, τ_h, of the atmosphere (1.0×10^7 s) is much greater than the planetary rotation rate, τ_r, (8.64×10^4 s)†. A rotation rate of this magnitude is, however, of considerable significance. Conservation of angular momentum results in an east–west component in atmospheric motion. On the Earth the effect of rotation is seen in the direction of the surface winds at all latitudes. The intense upper westerly winds (known as the jet stream) (figure 2.6) are a planetary scale phenomenon resulting from the pole-to-equator temperature gradient and the rotation rate, as was demonstrated by Hide (1965, 1980). Figure 2.6 illustrates the interaction of cellular convection and large-scale Rossby waves under the influence of these driving mechanisms for the atmospheres of Jupiter and the Earth. Differential heating combined with the rotational velocity of the planet produces

† cf Mars $\tau_h = 2.0 \times 10^5$ s and $\tau_r = 8.9 \times 10^4$ s.

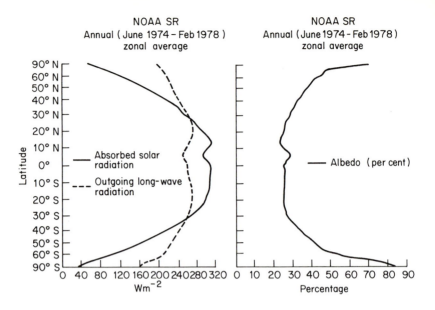

Figure 2.5. Annual averaged curves of latitudinal absorbed and emitted radiation. The absorbed solar radiation is a strong function of the albedo and decreases rapidly towards both poles. The infrared radiation emitted is a much weaker function of latitude indicating the efficiency of the atmospheric and hydrospheric meridional heat transport mechanisms on the Earth (after Winston *et al* 1979).

various types of flow in the modelled atmospheres. The features and the strength of the flow are found to be functions of the imposed temperature gradient as well as the speed of rotation. Hide (1980, 1981) has simulated individual zones in the atmosphere of Jupiter as well as the full planetary circulation, producing a configuration of closed eddies which have a stability and persistence analogous to that of the Great Red Spot and white ovals.

Direct comparison between the circulations of the atmospheres of Jupiter and the Earth may be unwarranted. This is because the energy supply for the Earth's atmospheric circulation is drawn directly from the pole-to-equator temperature gradient. At cloud-top level on Jupiter the temperature difference is about 3 K (cf Earth's value of about 30 K) and the pole-to-equator distance is approximately an order of magnitude larger. Furthermore, the solar energy source may well be much less significant than the internal heat source for Jupiter and Saturn (§3.1.1) cf the terrestrial planets.

The Earth's atmosphere is currently responsible for reducing the considerable latitudinal radiation imbalance (figure 2.5) and thus for establishing a more homogeneous temperature regime. Since the incorporation of gases into surface and subsurface reservoirs is a function of local surface temperature such a moderating effect is extremely important.

Although the atmospheres of Venus, the Earth and Mars are all driven directly by solar heating which produces a pole-to-equator temperature gradient, there are considerable

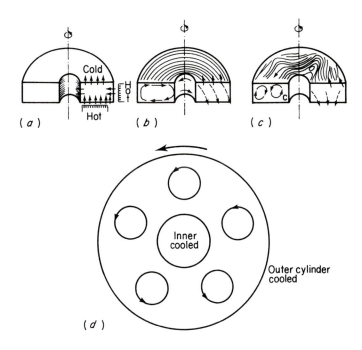

Figure 2.6. The rotating annulus experiments of Hide (1965, 1981). The heating distribution for diagrams (*b*) and (*c*) is shown in (*a*). At low apparatus rotation speeds (*b*) a Hadley cell (figure 2.5) circulation is set up—note the variation of flow direction with height in the fluid. A Rossby wave regime (*c*) ensues when the rotation speed is increased. The wave pattern moves in the same direction as the apparatus. When both inner and outer cylinder walls are cooled (*d*) a configuration of closed and apparently stable (Hide 1981) eddies is seen which may resemble Jupiter's closed circulating vortices.

differences even between these apparently similar systems. Venus and the Earth achieve the latitudinal radiation imbalance at a different level in their atmospheres (the surface of the Earth cf the cloud tops of Venus). The slowness of the planetary rotation on Venus (243 days, table 1.2) and the considerable opacity of the atmosphere seem to combine to give rise to the spiral circulatory system in which air moves from equatorial regions finally to descend over the poles. The atmospheric rotation period seems to be approximately 4 days. The apparently anomalous evolution may be associated with this unique dynamical system (see §3.4).

Circulation patterns at lower layers in the cytherean atmosphere are not well understood (Counselman *et al* 1980) but it seems that meridional transport is less important than relaxation to radiative equilibrium. The cloud morphology is found to be considerably more complex than was anticipated; for instance the 'Y' or 'V' shaped feature seen in the ultraviolet images of Venus appears not to be locked to the Sun. It is not, therefore, a divergence of flow around the subsolar point but, on the contrary, is observed to rotate around the planet once in 4–5 days (Rossow *et al* 1980).

Short-term climatic variability on Mars has been linked to the atmospheric dynamics. For example, the thermal structure of the whole planet–atmosphere system changes completely when dust particles are lifted into the atmosphere (Cutts *et al* 1979, Leovy and Zwek 1979). Hoffert *et al* (1981) have constructed a $2D$ energy balance model of the martian atmosphere which successfully simulates both the current and a 'fluvial' regime (chapter 5). Pollack (1979) has suggested that the laminated terrain around the north polar region is a result of differential terminal velocities of dust + H_2O + CO_2 particles compared with dust + H_2O or dust alone. The erosive nature of the tenuous atmosphere on Mars may also be critically dependent upon the wind velocity and dust carrying ability (Carr 1980, Spitzer 1980). Thus atmospheric dynamics control local-scale phenomena which in turn may affect climatic and evolutionary changes.

Some of the outer planets have internal heat sources (the emitted energy from Jupiter is more than 1.8 times greater than the incident solar radiation and even larger for Saturn, see chapter 3). Such an internal energy source reduces the effect of differential solar heating. The additional effect of very rapid rotation rates means that the atmospheric motions on these planets is latitudinal. Numerical simulations of Jupiter suggest that baroclinicity should characterise the atmosphere and thus that the zone and belt structure (unless driven by a large internal heat source) should be less well defined than observations reveal (Gierasch 1981)—currently the latitudinal velocities between the bright zones and dark belts are estimated to be about 100 ms^{-1} (§3.1). At deeper levels in the giant planets, internal phase changes (especially of He) could be responsible for evolutionary changes in the atmospheric composition.

It is important to be able to establish the temperature, pressure, density and chemical composition as a function of height in the atmosphere. As already discussed, the physical and chemical regime in the exosphere and at the surface will determine major changes. Photochemical reactions which may result in evolutionary changes in the atmosphere cannot be investigated in the absence of density and temperature data. Condensation of volatiles in the atmosphere and stability against surface forced convection are also functionally dependent upon the vertical structure.

2.1.4. Hydrostatic Control

In any atmosphere the compressible nature of the gases gives rise to pressure and density gradients with height. Assuming hydrostatic equilibrium, the barometric law

$$p(z) = p(z_0) \exp \left[(z - z_0)/H \right] \qquad (2.2)$$

and the complementary form in terms of density describe the variation with height in an isothermal atmosphere. The additional effect, upon both pressure and density, of temperature variations is shown for the case of the Earth in figure 2.7 together with the vertical temperature profile. Calculation of scale heights H, for all planetary atmospheres reveals a strong similarity. This seems to be coincidental. On Mars the low value of g is balanced by the high molecular mass of carbon dioxide, the predominant volatile, whilst for instance

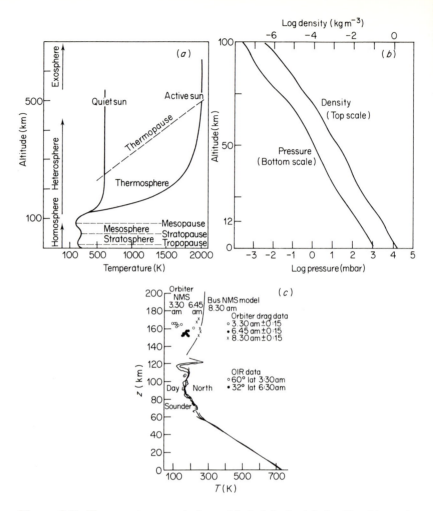

Figure 2.7. Temperature variation with height in (*a*) the Earth's and (*c*) Venus' atmosphere. (*a*) The regions of the Earth's atmosphere are defined by the thermal structure. (*b*) Pressure and density both decrease exponentially with height in the atmosphere. (*c*) The thermal structure of the atmosphere of Venus as sensed by the Pioneer Venus probes (after Seiff *et al* 1980) indicates a tropospheric lapse rate not very different from that of the Earth ($\sim 7\,\mathrm{K\,km^{-1}}$) but the surface temperature and upper atmospheric structure are different.

on Jupiter the low temperature and the high gravitational acceleration are counteracted by the very much lower molecular mass of the predominant constituent, hydrogen.

A temperature profile (from Pioneer Venus) of the cytherean atmosphere is also shown in figure 2.7 (after Seiff *et al*

1980). The surface temperature on Venus is approximately 730 K. Calculations suggest (Seiff *et al* 1980) that both the vertical and horizontal temperature fields can be understood primarily in terms of radiative equilibrium conditions. However since only about 17 W m^{-2} (or $2\frac{1}{2}\%$ of the incident radiation) is absorbed at the surface of the planet, it is not yet certain that the atmosphere possesses the required optical depth ($\tau \simeq 100$ suggested by Pollack *et al* 1980b). Furthermore the Pioneer Venus data suggest that the lapse rate in the lower atmosphere is slightly anadiabatic due to the presence of Hadley circulation (see e.g. figure 2.6). The warm polar region at the cloud-top level is probably a dynamical effect. The Pioneer Venus observations have considerably strengthened the arguments that the major component of the clouds which cover the planet is an aerosol of droplets of concentrated H_2SO_4 in solution†.

As already described, it is no longer believed necessary to invoke dynamics as any more than a secondary component in the explanation of the very high surface temperatures. However, Pioneer Venus has confirmed the existence of a Hadley cell circulation in the upper atmosphere of the planet (i.e. 50–80 km). Observations indicate that as well as the enormous zonal velocities (~ 100 m s^{-1}) there are meridional motions: poleward in both hemispheres of the order of 2–5 m s^{-1} at approximately 60–70 km and equatorward at around 50 km with speeds of about 5 m s^{-1} or less (Keating *et al* 1980). Over 50% of the incident solar flux is absorbed in this cloud layer but mean equator-to-pole temperature differences of only 10–20 K are observed. These observations (Taylor *et al* 1980) combine to demonstrate the existence of a Hadley cell circulation pattern.

A very tenuous atmosphere such as that on Mars today may behave more like a liquid than a gas if the atmospheric scale height, H, and the height of the topography are similar. This is because the atmosphere is then physically constrained by the presence of the larger features.

Above the well mixed region (troposphere) the barometric

† It is suggested that the optical properties (particularly the observed polarisation) are caused by nearly spherical drops with an effective radius of the order of 1.05 μm (Rossow *et al* 1980).

law should be applied to each individual gaseous constituent since the effects of gravity upon light elements (e.g. hydrogen and helium) is much less than upon oxygen, nitrogen, etc.

The escape of gases from planetary atmospheres is a function of the energy of individual molecules or atoms in the upper layers of the atmosphere. The most extreme form of escape is hydrodynamical flow or 'blow-off' which has been shown by Hunten (1973) to be important only when the scale height of the lightest atmospheric gas is greater than half the exobase height. It is then a very rapid process affecting all constituents. Forms of escape from the exosphere are: thermal escape, photochemical reactions leading to non-thermal escape (both discussed in §2.2) and escape through interaction with the solar wind. Hydrodynamic escape and Jeans escape are both forms of thermal escape mechanisms. Hydrodynamical escape is most applicable when the escaping constituent is the dominant gas at the exobase height. When the loss rate of a gas is derived in terms of the velocities of individual molecules the process is termed Jeans escape. Walker (1977) reviews these mechanisms. Lifetimes for individual elements on the planets are discussed below in terms of Jeans escape.

2.1.5. Magnetic Field Effects

The only important effect the existence of a planetary magnetic field has in terms of the evolution of the atmosphere is the protection it affords to charged particles in the upper atmosphere from direct interaction with the solar wind. For instance, on the Earth the impinging particles are deflected by the magnetic field though some are actually channelled by the field lines to the polar regions where interaction can take place. On planets without magnetic fields (e.g. Mars and Venus) there is a direct and continuing interaction between elements of the solar wind and charged atmospheric particles. However, the net effect of these interactions is not yet well established for either planet and may change as a function of the chemical composition and atmospheric mass. Walker (1977) describes how the ^{3}He budget of the Earth's upper atmosphere can be used to set upper limits on the influx of ^{4}He and H from the solar wind. Using a ^{3}He influx value of $7 \times 10^4 \, \text{m}^{-1}\text{s}^{-1}$ he establishes that additions from the solar wind for the Earth seem likely to be a minor influence on the evolutionary process. However since the composition of the solar wind is close to that of the Sun itself, it has been suggested that addition of volatiles directly from deposition by the solar wind could account for certain of the apparent discrepancies amongst the noble gas and volatile ratios on the planets (see tables 2.1(b) and 3.5). For instance, if all the solar wind currently incident at Venus were to be added to its atmosphere it has been calculated that the total accretion of hydrogen over the history of the solar system would be of the order 6×10^{29} atoms/m^2† (Pollack and Yung 1980). Volatile inventories as well as current flux measurements indicate that no such supply mechanism is likely to have occurred. It is interesting to note that a solar T-Tauri stage which is generally assumed to be responsible for the removal of atmospheric constituents (§1.2) could also be an important source of material under appropriate physical conditions. Cloutier *et al* (1969) underline the inability of the solar wind to remove unlimited amounts of atmospheric material. They calculate that if an increase of more than 56.25% by mass occurs in the upper atmosphere, a negative feedback mechanism resulting in a diversion of the flow is invoked.

An interesting effect occurs when the magnetic field interacts with a planetary satellite. The Galilean satellite Io is the trigger for the bursts of decametric radiation observed to come from Jupiter. The interaction between Io and Jupiter's magnetic field not only causes a current flow to Jupiter but must also induce considerable heating of the surface of Io itself (Hide 1980). This heating could be partially responsible for the volcanic activity observed on Io by the Voyager spacecraft (Carr *et al* 1979). Magnetic field studies of the Saturn system by Voyager 1 (Ness *et al* 1981) may indicate a similar source of energy for the satellite Titan. Titan has no significant magnetic field of its own (see chapter 3) and therefore direct interaction between the magnetic field of

† This falls between the upper and lower bounds of thermal escape of hydrogen from Venus of approximately 10^{33} and 10^{27} atoms/m^2 (Walker 1977).

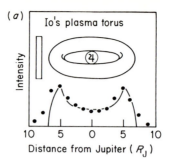

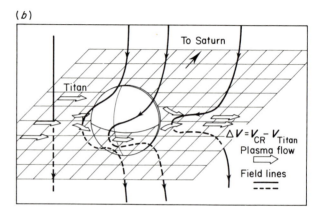

Figure 2.8. Both (*a*) Io and (*b*) Titan are close enough to their parent planet to interact with it in a number of ways. The magnetospheres of (*a*) Jupiter and of (*b*) Saturn extend to the orbital radius of these moons. The decametric radio outbursts from Jupiter are linked to electromagnetic interactions between Io and Jupiter. Furthermore both moons have toroidal-shaped plasma/gaseous constituents at their radial distance. Both may recycle some of their atmospheric constituents within this torus and possibly with the extended atmosphere of the parent planet ((*a*) after Broadfoot *et al* 1979 and (*b*) after Ness *et al* 1981).

Saturn and Titan's upper atmosphere takes place. Figure 2.8 illustrates the distortion caused in Saturn's field lines by Titan. Unlike the case of Io this interaction will be much less significant (possibly even negligible) at the surface since Titan possesses an atmosphere $1\frac{1}{2}$ times more massive than that of the Earth. It may however be an important perturbing influence for the upper layers of the atmosphere.

2.2. Chemical State and Loss of Gases

The chemical composition of the atmospheres around the early terrestrial planets has been the subject of debate in the literature for over a decade. Abelson (1966) favoured atmospheres containing mainly carbon dioxide and nitrogen and more recent geological (Schidlowski *et al* 1980a) and astronomical (Henderson-Sellers *et al* 1980) work has led to general agreement that gases emanating from the planetary surfaces and interiors are more likely to have been neutral or weakly reducing than highly reduced. Thus water will be more abundant than hydrogen, carbon dioxide more abundant than methane and nitrogen more abundant than ammonia (Walker 1978c). Indeed the proposition that the terrestrial planetary atmospheres could have been highly reducing is extremely difficult to substantiate, especially when one considers the necessity of planetary scale chemical upheavals taking place within all the planet–atmosphere systems since the current atmospheres are composed predominantly of CO_2, H_2O, N_2 (see table 1.4). There are, however, suggestions in the literature that planetary atmospheres could undergo considerable chemical changes. The models of, for instance, Holland (1962, 1978) and Fanale (1971) distinguish several stages determined by the oxidation state of the atmospheres. The earliest stage (see e.g. table 2.2, after Holland 1962) is characterised by reduced atmospheric gases which, it is argued, are the result of degassing activity prior to the formation of the metallic iron core.

Walker (1977, 1979) and Owen (1978), summarising the arguments of others, suggest that this migration of metallic iron would have been accompanied by a similar movement of nickel in the planetary interiors. Such a migration is not borne out by geological evidence and hence they suggest that core formation preceded all degassing activity of importance.

There is a second method of introducing reduced constituents into the atmospheres around the terrestrial planets. This is by collision with comets. The climatological effects of such an impact might well be considerable (Lazcano-Araujo and Oro 1981), but it is unlikely that such an impact could totally disrupt the neutral nature of the gases within the

Table 2.2. Postulated atmospheric evolutionary states and gaseous constituents (after Holland 1962).

	Stage 1	Stage 2	Stage 3
Major components ($P > 10^{-2}$ atm)	CH_4 H_2 (?)	N_2	N_2 O_2
Minor components ($10^{-4} < P < 10^{-2}$ atm)	H_2 (?) H_2O N_2 H_2S NH_3 Ar	H_2O CO_2 Ar	Ar H_2O CO_2
Trace components ($10^{-6} < P < 10^{-4}$ atm)	He	Ne He CH_4 NH_3 (?) SO_2 (?) H_2S (?)	Ne He CH_4 Kr

troposphere. It is both interesting and exciting to note that the origin and earliest evolutionary stages of life no longer appear to require strongly reduced conditions (Schwartz 1981). On the contrary, in chapter 4 it will be demonstrated that surface conditions very like those of the present-day Earth (but in the absence of free oxygen) are a more important requirement for this origin and evolution than gaseous chemicals in reduced form.

Removal of gases produced from the planetary system itself will be either through chemical and physical mechanisms leading to incorporation within the surface or through escape to space from the top of the atmosphere (as described in §2.1). Gaseous escape from planetary atmospheres is probably as strongly constrained by tropospheric parameters as it is determined by the characteristics of the exosphere. Hunten (1973) has shown that the rate of loss of hydrogen from the Earth is controlled by the upward diffusion rate of all hydrogen species from the lower atmosphere. The primary control is therefore the limit set on the upward movement of water vapour by the tropopause or 'cold trap' temperature.

2.2.1. Thermal Structure

The thermal structure of the atmosphere is critically important at all levels. At the surface the mean temperature and the temperature extremes govern the condensation and sublimation/evaporation of volatiles and the ambient temperature affects the rate of some chemical reactions. In the exosphere the temperature controls the velocities of individual particles and hence through the Maxwell–Boltzmann distribution the flux of each element from the planet can be calculated. For example since the most probable velocity V_0, is given by

$$V_0 = (2kT/Mm_H)^{1/2} \qquad (2.3)$$

where k is Boltzmann's constant, M the atomic weight to be multiplied by the mass of a hydrogen atom, m_H. The value of V_0 for hydrogen in the early Earth's exosphere ($T \simeq 600$ K, cf the present value of $\simeq 1500$ K) is about 3 km s^{-1} compared to the escape velocity at this height of about 11 km s^{-1}. From the Maxwell–Boltzmann distribution it can be shown that thermal escape could account for the loss of all free hydrogen from the Earth in considerably less than the 4.5×10^9 years of its life. Atomic oxygen has a much lower value of V_0 ($\simeq 0.8$ km s^{-1}) and therefore its escape is most unlikely through the thermal mechanism. Hunten (1973) has shown that the characteristic Jeans escape time for hydrogen from the present-day Earth is only a few hours which is currently shorter than the replenishment time by hydrogen species from below. Thus at any stage in the planetary evolutionary history exospheric escape rates may be dependent upon conditions lower down in the atmosphere.

The overall thermal structure of planetary atmospheres is dominated by the absorption of solar radiation, although in the case of some of the major planets there is a significant planetary component which must be considered as well. There are three layers (figure 2.7) of elevated temperature: the surface, close to the stratopause and within the thermosphere. It is at these levels in the atmosphere that significant absorption of solar radiation is taking place. The temperature within the troposphere is controlled by surface absorption of solar radiation and reradiation and convection from the planetary surface which then in turn warms the troposphere. The absorption process high in the thermosphere is photo-ionisation which is discussed in §2.2.2. The middle warm layer in the Earth's atmosphere is at about a height of 50 km and is the direct result of absorption of ultraviolet radiation by ozone. In the case of the evolution of the Earth's atmosphere the interaction between solar energy and the absorbing gases will have changed considerably. uv flux from the Sun (§1.2) and the existence of an ozone layer both change over time (chapter 4). Similarly the interaction between the ionosphere and the flux of charged particles from the Sun will have changed for all the planets.

A first-order calculation of temperature lapse rates can be performed by assuming *radiative equilibrium*. It should be noted that in most planetary atmospheres, convection and indeed the transport of energy through phase changes of volatiles will also be important mechanisms. However the assumption of radiative equilibrium (equation (2.1)) is a good starting approximation (albedo values for the planets are listed in table 2.3). Clearly either an increase in the planetary

Table 2.3. Albedos of the planets. Certain of the discrepancies arise from the date and method of estimating the albedos (see also table 3.4). Generally higher albedos are associated with substantial atmospheres (from Goody and Walker 1972, Allen 1963).

Planet	Albedo[a]	Albedo[b]
Mercury	0.058	0.059
Venus	0.71	0.85
Earth	0.30	0.40
Moon	0.07	0.068
Mars	0.17	0.15
Jupiter	0.73	0.58
Saturn	0.76	0.57
Uranus	0.93	0.80
Neptune	0.84	0.71
Pluto	0.14	0.15

[a] From Goody and Walker (1972) p 47.
[b] From Allen (1963) p 125.

albedo, or a decrease in the incident solar flux will result in a lowered value of the effective temperature, T_e. The difference between the planetary surface temperature T_s and the effective temperature is known as the 'greenhouse increment' such that

$$T_s = T_e + \Delta T \qquad (2.4)$$

where ΔT is the greenhouse increment which is caused by atmospheric absorption of outgoing thermal infrared radiation from the planet. (ΔT for the Earth now is approximately 33 K.) The greenhouse effect is caused by atmospheric gases which possess significant infrared absorption features. Figure 2.9 shows the Planck curves for the Sun (emitting temperature $\simeq 5700$ K) and the Earth (an emitting temperature of T_e approximately equal to 255 K, from equation (2.1)), together with the absorption curve of the Earth's atmosphere.

A zero-order approximation to the temperature profile of any atmosphere can be calculated using a grey approximation (i.e. the infrared absorption is averaged over wavelength). Using the assumption of local thermodynamic equilibrium for an optical thickness, τ, the radiative transfer equation gives

$$T^4(\tau) = T_e^4(1 + \tfrac{3}{2}\tau). \qquad (2.5)$$

At each level in the atmosphere radiation exchanges occur in upward and downward directions. However, at the surface this flux is unidirectional resulting in an equation for the surface temperature, T_s

$$T_s^4 = 2T_e^4(2 + \tfrac{3}{2}\tau_s) \qquad (2.6)$$

where τ_s is the wavelength-averaged optical thickness of the whole atmosphere. This gives a temperature discontinuity at the surface (figure 2.10). For large values of τ, the lower part of the atmosphere must become convectively unstable and dynamic adjustment will occur.

The effect of the atmosphere upon the thermal balance of the planetary surface is determined largely by the infrared properties of the individual gases. Table 2.4(a) lists the gases found in clean, dry air (for the Earth) at ground level together with the percentage of each, by volume. (The amount of water in the atmosphere varies considerably with

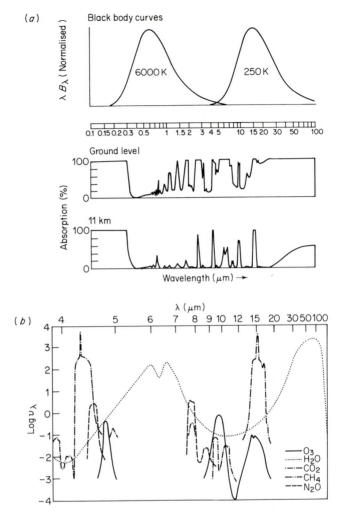

Figure 2.9. Normalised Planck curves of emitted energy for the Sun ($T_e \simeq 6000$ K) and the Earth ($T_e \simeq 250$ K). The atmospheric absorption at two levels in the Earth's atmosphere is shown in (*a*) (after Goody 1964). Tropospheric absorption predominates in near and thermal infrared regions whilst the effects of stratospheric ozone can be seen at shorter wavelengths. The specific molecular constituents responsible for the absorption features is illustrated in (*b*) (after Allen 1963).

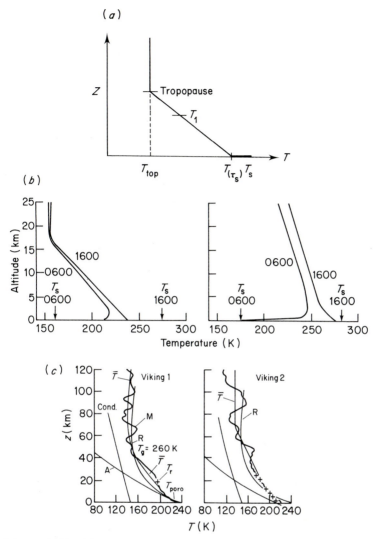

Figure 2.10. (*a*) A schematic diagram of the radiative equilibrium temperature structure may be compared with (*b*) the calculated temperature profiles for Mars from Gierasch and Goody (1972). The two curves show the importance of atmospheric dust in the Martian atmosphere. The dust loaded case is to the right-hand side. (*c*) Atmospheric temperature profiles retrieved by the Viking spacecraft (after Seiff and Kirk 1977) are also compared with the predicted curves of Gierasch and Goody (1972). Large temperature oscillations occur higher in the atmosphere. These are probably due to internal gravity waves forced by surface heating (Gierasch 1981). A, adiabatic profile; Cond, condensation boundary; M, measured profile; R, radiative equilibrium prediction.

latitude and season, but an average value is about 1 per cent of the total.) The very small mixing ratios of some of the gases listed in table 2.4(a) belie properties of importance. Table 2.4(b) is a summary of the mixing ratios and sources of the trace gases in the Earth's atmosphere. The infrared absorption spectrum (also typical of the Earth's atmosphere) shown in figure 2.9 illustrates the complex absorption features due to band absorption, not only by carbon dioxide and water vapour, but also by methane, ozone and nitrous oxide. These and other gases possessing infrared absorption bands (e.g. ammonia) will influence atmospheric and planetary surface evolution. Even apparently chemically inert gases, such as nitrogen and argon (see e.g. table 2.1(b)), cause broadening of the radiation absorption features and hence, if they exist in significant quantities, can modify surface

Table 2.4. (a) The chemical composition of the Earth's atmosphere from Schidlowski (1980b) in which sources are cited. (b) Origin of trace gases in the Earth's atmosphere (after Levine and Allario 1982).

(a)

Constituent	Formula	Abundance by volume ($\%$, ppm, ppb)	
Nitrogen	N_2	$78.084 \pm 0.004\%$	
Oxygen	O_2	$20.948 \pm 0.002\%$	
Argon	Ar	$0.934 \pm 0.001\%$	
Water vapour	H_2O	Variable ($\%$-ppm)	
Carbon dioxide	CO_2	335	ppm
Neon	Ne	18	ppm
Helium	He	5	ppm
Krypton	Kr	1	ppm
Xenon	Xe	0.08	ppm
Methane	CH_4	2	ppm
Hydrogen	H_2	0.5	ppm
Nitrous oxide	N_2O	0.3	ppm
Carbon monoxide	CO	0.05–0.2	ppm
Ozone	O_3	Variable (0.02–10	ppm)
Ammonia	NH_3	4	ppb
Nitrogen dioxide	NO_2	1	ppb
Sulphur dioxide	SO_2	1	ppb
Hydrogen sulphide	H_2S	0.05	ppb

Table 2.4. (*Continued*)

(b)

	Ground Concentration		Source		Atmosphere
	% Volume	Mixing Ratio[a]	Outgassing	Biological[b]	
Neon (Ne)	1.8×10^{-3}	18 ppm	Yes—radiogenic	No	No
Helium (He)	5.24×10^{-4}	5.24 ppm	Yes—radiogenic	No	Escapes
Methane (CH$_4$)	1.6×10^{-4}	1.6 ppm	No (?)	Yes	No
Krypton (Kr)	1.4×10^{-4}	1.14 ppm	Yes—radiogenic	No	No
Hydrogen (H$_2$)	5×10^{-5}	0.5 ppm	Yes	Yes	Chemical production
Nitrous Oxide (N$_2$O)	3.2×10^{-5}	0.32 ppm	No	Yes, Man	Lightning
Carbon Monoxide (CO)	1.2×10^{-5}	0.12 ppm	Yes	Yes, Man	Chemical production, lightning
Xenon (Xe)	8.7×10^{-6}	87 ppb	Yes—radiogenic	No	No
Ozone (O$_3$)	3.0×10^{-6}	30 ppb	No	No	Chemical production
Ammonia (NH$_3$)	1.0×10^{-7}	1 ppb	No	Yes, Man	No
Sulphur dioxide (SO$_2$)	2×10^{-8}	0.2 ppb	Yes	Yes, Man	No
Hydrogen Sulphide (H$_2$S)	2×10^{-8}	0.2 ppb	Yes	Yes	No
Nitrogen Oxides (NO + NO$_2$)	1×10^{-8}	0.1 ppb	No	Yes, Man	Lightning, chemical production

[a] ppm is parts per million by volume; ppb is parts per billion by volume.
[b] Man indicates an anthropogenic source in addition to a biological or microbiological source.

temperatures. The presence of an unknown amount of neutral gas having the ability to broaden the absorption features of other gases without itself being readily detected could be an important factor in surface temperature calculations.

The primary cause of extinction in the infrared is absorption by molecules. The frequency, ν, of absorption/emission of radiation is related to the energy difference by

$$\Delta E = h\nu. \tag{2.7}$$

Thus if large photon energies are available the electronic states of the molecules may be excited causing absorption/emission in the ultraviolet and visible regions. Changes in the vibrational energy of a molecule result from absorption/emission of radiation of frequencies typical of the near and middle infrared while rotational changes cause features in the far infrared. The type of activity of any species will be a function of the charge displacement of the molecule: no vibrational spectra are expected from the symmetric vibrations of N_2 and O_2 (the two most abundant constituents of the Earth's atmosphere). Similarly the carbon dioxide molecule, which is linear and symmetric, has no permanent dipole moment and thus only two of the three possible vibrational modes produce spectral features, as is illustrated below. There is no pure rotation absorption. The different types of transition interact to produce typical molecular spectra.

Symmetrical stretch mode

$$\longleftarrow O \longrightarrow C \longleftarrow O \longrightarrow$$

(No spectral activity)

Asymmetrical stretch mode

$$O \longrightarrow \longleftarrow C \qquad O \longrightarrow$$

(Band at $\simeq 4.3\ \mu$m)

Bending or degenerate mode $\uparrow$

$$O \qquad C \qquad O.$$
$$\downarrow \qquad \qquad \downarrow$$

(Band at $\simeq 15\ \mu$m)

The spectral position of individual absorption features and

the relationship to other features and to the emission curve of radiation is very important (figure 2.9). It is necessary to consider the position and width of each feature. The general method of approach is to construct a theoretical model of band absorption for the gas from representations of the line absorption and an assumed spacing (Goody 1964).

An alternative approach to these complex analytical models is a semiempirical approach using direct observations of band absorption. These models allow a restricted number of chosen parameters to vary with varying laboratory conditions and hence produce a useful representation of absorption under different conditions (e.g. Henderson-Sellers and Meadows 1977, 1979b). This method of calculation is an improvement on the grey atmosphere approximation and indeed the planetary temperatures derived seem to be a more satisfactory estimate of the variation of T_s with time (chapter 3); although mixing by convection within planetary atmospheres is clearly going to be of significance in any situation where the optical depth becomes large. It is possible to model convective mixing by adjusting radiatively calculated temperature profiles to a selected convective lapse rate and thus effecting upward movement of energy. In making such a convective adjustment it is, however, also important that the effect of condensation of any gases which are likely to release considerable amounts of latent heat to the atmosphere are included. For instance, in the case of the Earth the tropospheric lapse rate is between the saturated adiabatic lapse rate (which ranges from approximately 3 to $10\,\mathrm{K\,km^{-1}}$) and the dry adiabatic lapse rate ($9.8\,\mathrm{K\,km^{-1}}$); it has a value of between 5 and $7\,\mathrm{K\,km^{-1}}$. Planetary surface temperatures and possible escape mechanisms from atmospheres depend critically upon the vertical temperature structure. It is thus necessary to make an appropriate convective adjustment before calculations of long-term evolutionary trends of temperatures can be made.

It should be noted here that the convective environment will be restricted to the lower atmospheric layers (<10–15 km on the Earth) and that radiative equilibrium is only applicable to heights below approximately 70–80 km in the Earth's atmosphere (figure 2.7). In situations where the planetary

55

atmosphere is too thin to interact strongly with thermal radiation, heating mechanisms dependent upon photo-chemical and photoionisation processes become important.

There is likely to be considerable variation in the temperature profile with latitude in certain planetary atmospheres. Furthermore, variation of insolation, both as a function of time of year, and as a function of time of day, creates very different temperature regimes within the atmosphere at a given position on the planet's surface. For instance, on Mercury a very strong diurnal variation of surface temperature is seen, but the atmosphere and certainly the atmospheric effect on the surface temperature is negligible. Hence, although mean planetary temperatures are about 350–400 K, the diurnal surface temperature variation can be as great as 200–800 K. However, on Venus there is little temperature variation either from the equator to the pole or from day to night (see figure 2.7).

A more interesting case is that of Mars, on which the tenuous atmosphere does indeed contribute to surface heating through reradiation in the thermal infrared wavebands, but there is a very strong diurnal signal causing tidal waves in the atmosphere (Gierasch 1981). Theoretical temperature profiles calculated for the martian atmosphere are displayed in figure 2.10. These early calculations of the temperature profile have been shown to be very satisfactory by the Viking Lander missions to Mars during 1976 and 1977 (Seiff and Kirk 1977, Spitzer 1980).

Estimates of temperature profiles on Jupiter are dependent upon radiometric and infrared measurements, made either from the Earth or from satellite systems passing or orbiting the planet. Figure 2.11 shows a schematic estimate of the temperature structure of the jovian atmosphere. It should be compared with Voyager measurements (figures 3.2 and 3.3). The estimated adiabatic lapse rate is of the order of 3 K km^{-1}. However, it is believed that there is considerable absorption by NH_3, solid CH_4 and NH_3 within the jovian atmosphere which accounts for the discrepancy between observed and calculated lapse rates. The particular properties and absorption bands of these reducing gases are mentioned in §3.1.

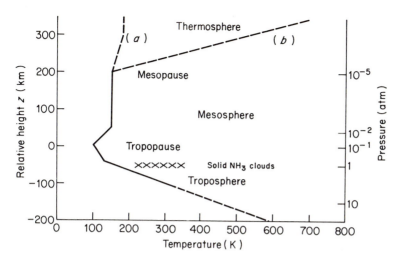

Figure 2.11. Theoretical temperature structure of the jovian atmosphere (after Chamberlain 1978). The value of the adiabatic lapse rate in the troposphere is uncertain, since it depends on the H_2/He mixture and the role of moist convection. The tropopause probably occurs at a minimum T of 100–120 K. A stratosphere is not shown because the profile is not known accurately enough for one to be sure that there is a temperature decrease above the tropopause. In the region just above the tropopause, T probably increases because of direct solar absorption, with a low level of infrared activity providing radiative cooling. At the mesopause the main vibrational relaxation occurs in C_2H_2 and CH_4. In the thermosphere, curve (a) is the expected profile for heating by solar extreme ultraviolet and curve (b) is a profile inferred from the Pioneer 10 radio occultation. Compare this theoretical temperature profile with the observational results shown in figure 3.3.

2.2.2. *Photochemical and Chemical Reactions*

Chemical reactions are one of the primary sources and sinks of gases in planetary atmospheres. The physical and thermal properties define the planet–atmosphere environment. The extent, location and rate of atmospheric chemical reactions are themselves a function of the surface temperature and the atmospheric temperature profile. The rate of chemical reactions is generally comparatively slow compared with reactions initiated as a result of the absorption of solar photons. The highly reactive free radicals produced as a result of solar

57

photon absorption initiate complex, and usually rapid, chemical reaction series in the troposphere and stratosphere. In our consideration of planetary evolutionary climatology, many chemical and photochemical processes are almost instantaneous. For instance, a number of authors have suggested that early surface temperatures on the Earth could have been increased by a considerable mixing ratio of ammonia. Kuhn and Atreya (1979) have investigated the lifetime of atmospheric mixing ratios of NH_3 great enough to produce a significant effect in the early atmosphere. They suggest that while a mixing ratio of 10^{-5} NH_3 produces ΔT between 12 and 15 K, it has a lifetime of less than 10 years against photodissociation; a mixing ratio of 10^{-4} of NH_3 is completely removed by photodissociation in less than 40 years (see chapter 4).

Photochemical and photoionisation reactions are initiated by direct absorption of energetic solar radiation by a molecule or atom, the energy of each photon being inversely proportional to the wavelength of the radiation. Whilst the mechanisms are similar on all the planets, the nature and rate of reactions is a function of the relative abundances of gases in the upper atmospheres. On Venus the majority of the reactions are between high-energy photons and carbon dioxide molecules although at heights above 150–160 km atomic O becomes the most abundant constituent in the cytherean atmosphere. On the Earth the predominant reaction is photoionisation of O to O^+. Production of excited atoms leads to further chemical reactions (see e.g. Goody and Walker 1972, Walker 1979).

The three major types of photochemical reactions which may result in removal of atoms from the atmosphere of Mars are described by McElroy *et al* (1976). They cite the following example reactions:

I   dissociative recombination:
$$O_2^+ + e \rightarrow O + O$$

II   photodissociation:
$$CO + h\nu \rightarrow C + O \tag{2.8}$$

III   collision with photoelectrons leading to dissociation:
$$e + N_2 \rightarrow N + N + e.$$

It appears that photochemical mechanisms could provide substantial sinks for C, N and O on Mars by providing atoms or ions with enough energy to escape. Yung and Pinto (1978) have given further support to the view that an atmosphere composed of reduced gases is unlikely to persist even if formed, by demonstrating that a 1000 mbar atmosphere of CH_4 on Mars is likely to be lost through photochemical conversion to higher hydrocarbons in approximately 10^8 years. The heavier, more complex molecules would gravitate to the surface and be lost to the atmosphere.

The maximum ozone density in the Earth's atmosphere occurs at about 25 km although there is considerable seasonal variation (figure 2.12). Photolysis of ozone molecules by ultraviolet radiation starts at approximately $11\,000 \times 10^{-10}$ m; maximum absorption occurring between 2000 and 3000×10^{-10} m:

$$O_3 + h\nu \rightarrow O_2 + O. \tag{2.9}$$

The sequence of processes responsible for the absorption of

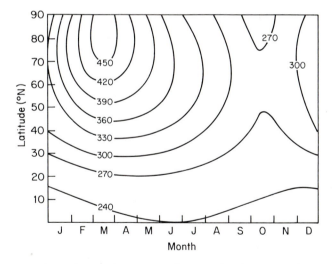

Figure 2.12. The total amount of ozone in the atmosphere as a function of month and of latitude, averaged over a period of several years in the northern hemisphere. The lines connect points of equal ozone amount. The units give the thickness, in thousandths of a centimetre, of a layer composed of all of the ozone in the atmosphere compressed to 1 atmosphere pressure (after Godson 1960).

high-energy ultraviolet solar radiation, the removal of ozone, and finally the recombination of an oxygen atom and an oxygen molecule, in the presence of a third body, to reform ozone molecules at this particular height in the atmosphere is named the Chapman process. Stratospheric photochemistry involving ozone has recently become of interest following suggestions that anthropogenic changes could result in destruction of O_3 and hence an increase in the UV flux at the Earth's surface.

The theories of photochemistry should be applied directly to the study of evolutionary histories of planetary atmospheres. Very little progress has yet been made in this area. Walker (1977) gives an excellent review of the type of analysis that should be undertaken. However the necessary framework for such photochemical studies does not yet exist. Since most of the reaction chains occur very rapidly and are extremely sensitive to temperature, pressure and chemical state, any such study can only be undertaken after the physical and climatological characteristics of the planetary environment have been established. Kasting and Walker (1981) state that photochemical model predictions are dependent upon 'the temperature of the primordial tropopause'. Walker (1977) undertakes an interesting and detailed examination of the photochemical reactions required to maintain an NH_3 mixing ratio greater than 10^{-6} in the early Earth's atmosphere because this value is 'required to maintain surface temperatures above freezing'. This premise is based on the work of Sagan and Mullen (1972) and, as described in chapter 4, is probably unsound because inappropriate (present-day) values of the physical characteristics of the Earth were used. In this work, photochemical reactions will only be considered if they can be shown to occur slowly enough that their timescale is of the same order as the evolutionary timescale. All other chemical reactions will be treated as occurring in negligibly short time periods† so that only their final outcome (i.e. the equilibrium situation) will be of concern here. This, in a sense, is a complementary approach to that of Walker (1977).

† This assumption seems to be reasonable for most elements on most of the planets. However Titan may prove to be an exception, see §2.5.3.

There exist three pathways for the removal of gases from planetary atmospheres: atmospheric reaction, loss at the lower boundary of the atmosphere (surface) or loss at the upper boundary (exosphere). These three potential sinks are not entirely divorced from one another, since, as has been seen, the availability of chemical species in the upper levels of any atmosphere is a direct result of composition and reactions lower down. For example, for water vapour it has recently been noted that the efficiency of the 'cold trap' is apparently smaller than had been suggested (Rind 1981). Data relating to the variation of water vapour amount in the region of the tropical tropopause indicate that the level of H_2O vapour is greater than expected. This suggests that water vapour can penetrate into the stratosphere even in the presence of a considerable temperature inversion (figure 2.7) although the mixing ratio is still very low compared with the steam atmosphere postulated for Venus by Ingersoll (1969). This effect may be of considerable significance for the evolution of any planet upon which a global hydrosphere is readily attained (chapters 4 and 6). The method, extent and rate of loss of atoms from the exosphere may be controlled by different features of this complex system at various stages in the evolutionary histories of the planets. For example, on Venus the atmospheric budget of helium is difficult to balance. This is due primarily to the sensitivity of the Jeans escape flux to exospheric temperature. Macdonald (1964) estimated that the average escape flux is about 30 times smaller than the surface source. A solution to this imbalance may be that a photochemical escape mechanism (cf earlier discussions of Mars) may operate.

Removal of gases from the troposphere to the surface sink is functionally dependent upon physical constraints as well as chemical reactions. The case of carbon dioxide in the Earth's atmosphere illustrates the extreme complexity of these systems. Early condensation of surface liquid water on the Earth permitted carbon dioxide to go into solution and form sedimentary rocks through both biological and abiological deposition processes in the marine environment (see chapter 4). Table 2.5(b) lists the Earth's carbon inventory compared with the total volatile inventory of the Earth (table 2.5(a))

Table 2.5. (*a*) Total outgassed volatile inventory for the Earth. (*b*) Inventory of carbon near the Earth's surface (normalised with respect to the biospheric values, see e.g. Bolin *et al* 1979). The biosphere and hydrosphere have removed the majority of the Earth's atmospheric CO_2 to these surface and subsurface reservoirs.

(*a*)

Volatile	Mass (10^{17} kg)
Water vapour (H_2O)	16 000
Carbon dioxide (CO_2)	910
Atomic chlorine (Cl)	300
Atomic nitrogen (N)	42
Atomic sulphur (S)	22
Atomic hydrogen (H)	10

(*b*)

Carbon reservoir	Normalised amount
Biosphere marine	1
non-marine	1
Atmosphere (in CO_2)	70
Ocean (in dissolved CO_2)	4 000
Fossil fuels	900
Shales	850 000
Carbonate rocks	2 000 000

illustrating the huge scale of this processing. Very much shorter timescale processing of atmospheric gases occurs both as a function of life systems and surface temperatures. Figure 2.13 shows the trends of CO_2 observed at Mauna Loa and at the South Pole (after Keeling *et al* 1976a,b). The oscillations in the curve are a direct result of the seasonal cycle biospheric uptake, superposed upon the generally increasing CO_2 levels resulting from industrial activity†. Schneider's (1975) review of the modern carbon dioxide greenhouse problem (Bolin *et al* 1979) underlines the complex feedback

---

† It is generally agreed that the likely global temperature increase resulting from a doubling of the CO_2 level is approximately 2 K (GARP 1975); a more recent estimate from General Circulation Models (GCM) is 3 K $\pm$ 1.5 K (Smagorinski 1981).

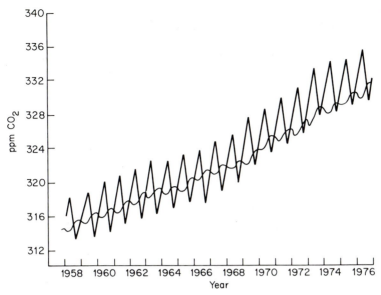

Figure 2.13. Increasing levels of CO_2 in the Earth's atmosphere as observed by Keeling *et al* (1976a,b) and Rycroft (1982) at Mauna Loa and at the South Pole. Superimposed upon the upward trend, which is clearly global in nature, is a seasonal oscillation which is controlled by the biosphere.

mechanisms linking the physical, chemical and temperature characteristics of the atmosphere.

2.3. Feedback Effects within Evolving Atmospheres

The major mechanisms allowing escape of gases from planetary atmospheres have been listed above. They are exospheric escape, condensation on to the surface (or into clouds—see below) and chemical reactions between atmospheric gases and the surface. The possibility of chemical reactions taking place will depend both upon the surface temperature and the partial pressure of the gas concerned and also upon the planetary surface. Over long periods the loss of volatiles through combination with the surface may be very important. For instance it has been suggested by Bolin *et al* (1979) that buffering of the carbon dioxide in the Earth's atmosphere by weathering reactions may be an important effect in temperature calculations (see §2.2.2). On Mars the removal of water vapour by photodissociation and the escape of hydrogen will only continue if the remaining free oxygen is removed—this will probably be by reaction with the surface (Levine *et al* 1978). Similarly chemical reactions at the surface of Venus may affect the overall atmospheric constituents and if a massive 'steam' atmosphere ever existed (Ingersoll 1969) it must have been lost either by escape or removal (Walker 1979). Losses are functionally dependent upon the vertical structure of the atmosphere (§§2.2.2 and 2.3.2). Prediction of the likelihood of condensation of volatiles on to the surface requires more than the knowledge of the mean global value of T_s. Condensation and hence cloud condensation (both solid and liquid) will depend upon the vertical distribution of the volatile and the temperature structure. Temperature structures determined in terms of radiation alone are unrealistic. The comparison of computed values of T_s and T_{top} (assumed to be congruent with the temperature of an isothermal stratosphere) for any model atmosphere will give an approximate description of temperature variation in the troposphere of the planet, if the chemical composition remains constant with height. From this temperature distribution a simple consideration of cloud formation may be possible (Weare and Snell 1974). This simple cloud condensation model would be an interesting development for planetary climatic modelling (Henderson-Sellers and Henderson-Sellers 1975). The estimation of changing cloud amount will be discussed in §2.3.2 and in chapter 5. In terms of the feedback effects on temperatures it seems likely that the loss of vapour/gas by condensation will produce a much greater variation in the values computed for the surface temperature than losses from the top of the atmosphere. Both will cause a change in the planetary albedo. Cloud formation (see §2.3.2) may also add to the greenhouse temperature increment.

2.3.1. *Surface Condensation*

Condensation of an atmospheric volatile on to the planetary surface will depend upon the lowest local temperature (or cold trap temperature) which will usually be considerably lower than the average global temperature, T_s. Under certain circumstances it can be argued that the extent of the atmosphere is a function of the mean surface temperature. A planet which has always had T_s below 270 K is probably not glaciated, but rather only partially degassed, since water vapour (and, for much lower values of T_s, carbon dioxide also) will not escape from the surface layers. This is clearly not true if degassing occurs primarily from volcanic emanations or as a result of impacts. Such a 'sub-surface glaciation' will evolve into a more hospitable environment as the surface temperature increases. The picture simplifies the slow degassing evolutionary sequence by requiring that the non-degassed volatiles take little or no part in the feedback loops (i.e. in albedo modification or surface partial pressures). Once the surface temperatures rise high enough to liberate large amounts of carbon dioxide and, later water vapour, then this liberation, probably over large areas of the planetary surface, will rapidly modify the average surface temperature by infrared absorption. The surface condensation of volatiles will be determined by the new average and local temperatures, but the partial vapour pressure remaining (and thus modifying the temperature further) will itself be a function of the amount of condensation which takes place. The feedback loops are seen to be highly complex. However, over long time periods (i.e. of the same order as the lifetime of the planet) many, if not all, of these rapid oscillations can be ignored, so revealing the general temperature trends. (Shorter term changes are discussed in chapter 5.) The present-day partial pressure of both carbon dioxide and water vapour on Mars is believed to be buffered by the polar caps and also by sub-surface ice deposits (see e.g. Leighton and Murray 1966, French and Gierasch 1979).

The surface temperature curves in chapter 3 are derived assuming a volatile to be absent (or almost absent) until the temperature is high enough to liberate significant amounts from the sub-surface layers. Once there is an established partial pressure of a volatile, it is possible that some may condense on to the surface (probably initially in polar regions), altering the albedo and the surface infrared emissivity. It is important to note that even if the global average temperature is suitable for condensation of a volatile, topography and local temperature variations will limit the areas of condensation (as is the case for liquid water on the Earth and carbon dioxide on Mars.) Thus the likelihood of a totally glaciated planet developing seems small (see chapters 4 and 5).

Cloud formation is a function of both the temperature profile of the atmosphere and the vertical distribution of the volatiles. The effect of clouds upon planetary climatic change may be of the utmost importance, as discussed below.

2.3.2. Clouds

As with many of the other arguments presented here, direct analysis and use of present-day terminology may lead to confusion. For instance, the troposphere derives its name from tropos meaning turning and the tropopause is the position of temperature gradient discontinuity. Use of these terms therefore suggests a well mixed lower atmosphere (i.e. by convection) and a temperature inversion above. In the absence of a stratospheric ozone layer the Earth's tropopause is likely to be considerably less well defined and even the height of the tropopause may be difficult to establish (see chapter 4). In the following discussion of the Earth the present value of approximately 10 km is assumed. This seems reasonable in the light of our discussion of likely atmospheric mass but the total depth must also be related to the strength of the convective mixing. Most important radiative processes take place within planetary tropospheres. Defining the tropopause temperature, T_{top}, as before (see figure 2.10) then T_{top} is the skin temperature of the model atmosphere. Thus following Goody and Walker (1972)

$$T_{top} = T_1 2^{-1/4} \tag{2.10}$$

where T_1 is the temperature at $\tau \sim \frac{1}{2}$. (For a non-grey model T_{top} is lower than this value.) Assuming a grey atmosphere solution of the radiative transfer equation (Goody 1964) gives

$$T^4(\tau) = (F/2\sigma)(1 + \tfrac{3}{2}\tau) \tag{2.11}$$

where τ is the optical depth of the atmosphere and F the flux. Thus

$$T_1 = T_e(8/7)^{-1/4} \tag{2.12}$$

and hence

$$T_{\mathrm{top}} = T_e 2^{-1/4}(8/7)^{-1/4}. \qquad (2.13)$$

Thus T_{top} appears to be solely a function of the effective temperature (equation (2.1)).

The calculation of mean global surface temperature described here presumes radiative equilibrium, i.e. no account is taken of the effect of convective processes within the atmosphere although any discussion of cloud cover is incompatible with this assumption. Inclusion of any convective adjustment should involve calculation or parametrisation of vertical energy transport which is currently extremely difficult. General circulation models of the atmosphere assume that an atmospheric lapse rate more unstable than $6.5\ \mathrm{Kkm}^{-1}$ is highly unlikely e.g. Manabe and Wetherald (1967) and a convective adjustment to this critical lapse rate is made. (GARP (1975) recommend that the method of parametrisation of the radiative–convective adjustment warrants further consideration.)

The radiative equilibrium lapse rates resulting from various model calculations are shown in table 2.6. (Calculations have been made for solar flux at 4.0×10^9 years BP and a global albedo of 0.3.) Table 2.6 suggests the conservative nature of the Earth's evolving troposphere, namely for the most likely atmospheric masses and absorber concentrations even the radiative equilibrium lapse rate does not depart too far from present-day environmental lapse rates. Any enhancement of surface temperatures through the addition of infrared absorbers probably leads to greater atmospheric instability. The possible resultant change in T_s and T_{top} can only be calculated if further assumptions are made about the nature of the convective activity. The practice used in general circulation models of constraining the environmental lapse rate to be less unstable than $6.5\ \mathrm{Kkm}^{-1}$ has been followed here. This is consistent with the evidence that for ambient surface temperatures close to the present day, extensive surface liquid water, surface pressures and water vapour mixing ratios have not differed greatly from the present-day values throughout the Earth's evolution (chapter 4). Table 2.6 also lists the resulting T_s and T_{top} values with this additional constraint. The

Table 2.6. Calculated values of $T_s(K)$ and the tropospheric lapse rate ($K\ km^{-1}$) for these likely atmospheric configurations. The results from the radiative and radiative plus convective equilibrium models are given for $A = 0.30$ and incident solar flux of $S = 1100\ W\ m^{-2}$ ($\sim 4.0 \times 10^9$ years BP), $T_{top} = 202.98$ K for the radiative case and 65 K less than T_s for the convectively adjusted atmosphere (after Henderson-Sellers 1981).

Atmospheric pressure (mbar)			Radiative		Convective	
CO_2	H_2O	Neutral	T_s	Lapse rate	T_s	Lapse rate
18.0	1.0	1000.0	324.10	12.11	267.62	7.14
70.0	1.0	1000.0	356.50	15.35	275.06	8.65
310.0	1.0	1000.0	360.82	15.78	275.62	8.79

possible environmental response to super-adiabatic lapse rates in terms of increased convective cloud formation is developed more fully in chapters 4 and 5.

These results suggest that any likely atmospheric configuration results in a convectively unstable troposphere. Thus it seems reasonable to assume (as indeed we have done implicitly throughout) that mixing, via convection, and cloud condensation can, and do, take place within the Earth's atmosphere (chapter 4). Cloud formation has the immediate consequence of perturbing the global albedo and hence directly affects T_{top} and this in turn perturbs T_s. The magnitude of this albedo feedback has been discussed by Henderson-Sellers (1979) and it appears that cloud variations may provide a negative (or stabilising) feedback effect in planetary atmospheric evolution. In chapters 4, 5 and 6 these processes are described more fully. However it is interesting to note here that the changes in the radiative properties and the release of latent heat† as a result of the condensation of an atmospheric volatile seem to provide constraints upon the planetary environment.

The simple cloud albedo feedback argument presented here neglects the effect of the vertical energy transfer itself. The convection described would reduce the surface tempera-

† Rossow (1978) has remarked upon the efficiency of the removal processes which govern the amount of water vapour in the Earth's troposphere and keeps the mean relative humidity slightly below unity.

ture and finally result in an increased tropopause temperature. Recalculation using the radiative transfer model but holding $T_{top} = T_s - 65$ K results in $T_s = 296.4$ K for a model atmosphere of 300 mbar CO_2 and 1 mbar H_2O from table 2.6 above. This recalculation has important consequences, namely: (i) the surface temperature is reduced; (ii) the tropopause temperature is increased; and (iii) the cycle of convective activity and hence cloud formation has been stabilised—as a result of model restrictions. If this proposed convective adjustment is considered preferable to the radiative equilibrium calculations then the discussion of average surface temperature given above should include the additional reduction due to convection. If geological data indicating considerably higher global surface temperatures than those of the present day are substantiated, the likelihood of more vigorous convective activity in the past will be increased. (However it should be noted that the temperature curves deduced by Knauth and Epstein (1976) have been subject to re-interpretation. Walker (1982) sugests that their interpretation underestimates both secular changes in the isotopic composition of sea water and in the cherts themselves.)

Feedback effects in the Earth's atmosphere as a result of cloud formation are extremely difficult to predict. For instance, it has been calculated (Roads 1978) that, contrary to the widely held view, increasing surface temperature leads to decreased cloud cover (as a percentage of the surface area covered). Sellers (1976) had previously suggested that such a decrease could be caused by the build up of convective clouds rather than stratiform clouds (see also Schneider 1972, Cess 1976). Roads (1978) suggests that the cloud amount will probably decrease by 1% per degree Kelvin increase in T_s. It is now possible to investigate the atmospheric response to changes in the Earth's surface temperature (within the ranges expected from the geological evidence). Here very large changes in T_s (± 10 K) (compare with the glacial/interglacial change of approximately 5 K) are considered in order to illustrate the stability even in extreme conditions, but the results also hold for smaller temperature changes. If T_s were to be increased from its present value of 288 K to 298 K the resulting increased convection would lead to a decrease in the

Table 2.7. Earth's albedo values, A, calculated for varying percentage cloud cover (after Henderson-Sellers 1979).

Percentage cloud cover	Albedo (%)
50 (present day)	30.1
40	24.6
60	35.7

cloud amount of around 10% and to a lower value of A (table 2.7). The new value of T_{top} is 211 K. Thus the lapse rate increases, compared with the present-day value of 7–8 K km^{-1} (Schneider *et al* 1978). This, in turn, is likely (on a global scale) to enhance cloud formation (negative feedback). Similarly $T_s = 278$ K increases the cloud cover and increases A to about 35.7% (see table 2.7) and gives $T_{top} = 202$ K. The resulting lapse rate decrease would tend to inhibit further cloud build-up (of all types) and thus secure a return to a stable regime. This globally averaged argument follows similar earlier discussions regarding the average atmospheric stability and the likelihood of cloud condensation (e.g. Goody and Walker 1972). It appears that even large changes in T_s are unlikely to produce any catastrophic responses from the atmosphere through cloud cover changes.

Cloud configuration is a function of the general circulation and may also be dependent on the predominant underlying surface type (Henderson-Sellers 1978, 1979). Over the time span of evolving planetary atmospheres such perturbations will also be important (chapters 4 and 5). Temperature has been identified as a fundamental parameter for planetary atmospheric evolution. Feedback effects within the atmosphere will depend upon the temperature lapse rate and/or the surface temperature. Certain physical properties (the rotation rate, atmospheric mass and chemical composition, albedo, cloud type and height) of planets have been shown to be extremely important for the determination of the atmospheric state. The mechanisms and relationships described are directly employed in the detailed discussions of specific planetary atmospheric histories in chapter 3.

Frontispiece, chapter 3. One of the erupting volcanos on Io discovered by Voyager 1 during its encounter with Jupiter on the 4th and 5th of March 1979. The picture was taken on the 4th of March at about 5.00 pm from a range of about half a million kilometres showing an eruption region on the horizon. A higher ultraviolet reflectivity in part of the erupting plume may be due to scattered light from very fine particles (the same effect which makes smoke appear bluish). Gases and particulate material are being propelled from the interior of this satellite to the surface to form a tenuous, and possibly transient, atmosphere. (Photography courtesy of NASA/JPL.)

3. Atmospheric Evolution in the Solar System

The atmospheric state at any stage is the result of the net fluxes of mass and energy. The two primary sites for addition and removal of atmospheric constituents are the surface and the exosphere. The physical and chemical processes which control interactions at these critical locations and throughout the atmosphere have been described in chapter 2. The major planets (i.e. Jupiter, Saturn, Uranus and Neptune) have undergone many significant atmospheric modifications. However, their large masses make escape negligible and the lack of surface processes reduces modifications to chemical recycling although there may be minor additions from space. At the other extreme the planets and satellites which possess very small atmospheres (e.g. Mercury, the Moon, Pluto) may have suffered atmospheric modifications but the significance for the planet is very minor because of the thinness of the atmosphere. The category of parent body which is most interesting is clearly one which, whilst being large enough to retain a significant atmosphere, is also small enough to have suffered modifications to the atmospheric constituents through surface processes and upper atmospheric fluxes. A discussion of the evolutionary histories of this group which includes Venus, Mars, Titan and possibly also Triton will comprise the main part of this chapter. The Earth is also a member of this group but the evolution of its atmosphere is reserved for chapter 4 for two reasons: firstly the data set for the Earth is much more extensive than that for any other system discussed, and secondly the atmosphere of the Earth has been more strongly modified by the existence of life than by any other single factor. The possibility of life originating and thriving elsewhere within and beyond the solar system is considered in chapter 6.

'Snapshots' of two contrasting atmospheres seen, for example, in Jupiter (see the frontispiece) and the Earth demonstrate the shortest periods of motion as well as illustrating some of the fundamental factors underlying the characteristics of these two contrasting systems. Chapter 3 considers the evolution of the jovian (§3.1) and terrestrial (§§3.2–3.4) planetary atmospheres. It is generally assumed that whilst evolutionary processes may have changed the atmospheres of all the terrestrial planets, the atmospheres of the giant planets are unlikely to have been subject to more than very minor variations. It may be interesting to examine the corollary of this hypothesis: namely that other pairs of 'snapshots' from different epochs in the evolutionary history of the solar system would show a constant Jupiter contrasting with a changing Earth.

3.1. Atmospheres of the Major Planets

Even though it is widely assumed that the atmospheres of the major planets have undergone little or no change since their origin, it is worthwhile to consider their present structure and state and to examine their probable histories. Certain possible mechanisms for change can readily be observed—for instance, the photochemical and thermochemical processes producing the organic compounds which colour Jupiter and Saturn. Lightning is observed on Jupiter† and a wide range of non-equilibrium species has been identified. The internal and external agencies, e.g. solar luminosity, internal gravitational energy, responsible for cloud formation may have changed throughout the lifetime of the solar system. Interaction with solar and cosmic charged particles and with the planet's own satellites may be the source of perturbations in the present-day upper atmospheres (e.g. figure 2.8) and are therefore likely to have caused at least this level of modification

† It is unlikely that lightning could ever be observed in Saturn's atmosphere because the skies are almost permanently illuminated due to the presence of the rings.

continuously throughout the history of the planets (Trafton 1977). Voyager observations of Jupiter and Saturn indicate that these two planets are massive enough to cause their own atmospheric evolution. Precipitation of He into the core of Saturn is probably responsible for the larger internal energy radiated and hence for significant atmospheric differences.

The jovian planets have a number of common features: (i) their orbital positions are beyond the asteroid belt; (ii) their densities are extremely low (table 1.2); and (iii) their major atmospheric constituent is probably the lightest gas, hydrogen. Helium has not yet been identified on either Uranus or Neptune. Williams (1975) suggests that the outer planets should be grouped into two pairs by distance with Pluto forming a third group. Certainly there are as many differences amongst the atmospheres of the major planets as amongst those of the terrestrial planets. Currently the grouping noted by Williams (1975) is exaggerated by the considerable difference between the states of data. The atmospheres of Jupiter and Saturn have been probed by Voyagers 1 and 2 whilst information relating to Uranus, Neptune and Pluto is primarily from surface observations. Our knowledge of the most distant planet, Pluto, is extremely limited. It has even been argued that Pluto is an escaped satellite of Neptune or Uranus rather than a planet formed in solar orbit. Indeed its orbital path is so highly eccentric that periodically it crosses that of Neptune, thus relinquishing its position as the most distant body in the solar system (Harrington and Van Flandern 1979).

It is likely that another common feature of these giant planets is a central metallic silicate core, although this inference is drawn as much from models of planetary formation as from observational data. The gaseous envelopes are known to be composed of approximately solar elemental material in the case of Jupiter and Saturn and a similar composition is generally assumed for the atmosphere of Uranus and Neptune. The atmospheres constitute 95%, 85%, 20% and 15% of the masses of Jupiter, Saturn, Uranus and Neptune, respectively (Podolak and Cameron 1974). The mechanism of planetary formation is still the subject of much debate in the literature (see chapter 1), but most models suggest that

74

gravitational forces caused a hydrodynamical collapse of captured primitive solar nebula material over a comparatively short time during the formation of the solar system (see e.g. Pollack and Yung 1980). The very close chemical resemblance of the jovian atmospheres (table 1.4) to this primitive nebula material suggests that gross evolutionary processes have been unimportant for the giant planets and certainly the traumatic chemical and physical upheavals which seem to have occurred around the terrestrial planets do not form part of the histories of the giant planets. Despite the fact that these massive atmospheres are undeniably stable, the dynamic, thermodynamic, photochemical, magnetic and electrical activity witnessed by the Pioneer and Voyager spacecraft have provided data and stimulated new interest in the mechanisms of jovian planetary atmospheric modification.

3.1.1. *Jupiter and Saturn*

By analogy with the Earth and the terrestrial planets, Jupiter is traditionally considered as the prototype for the jovian planets. It has a diameter 11.2 times that of the Earth and its mass is $2\frac{1}{2}$ times greater than that of all the other planets combined (tables 1.1 and 1.2); it is the second most massive body in the solar system. It rotates on its axis in a period of less than 10 hours and latitudinal velocities close to the boundaries between the bright zones (areas of high cloud) and the dark belts (lower atmospheric levels) (see the frontispiece and figures 3.1 and 3.2) are known to exceed 100 ms^{-1}. The low density of the planet suggests a 'primordial' composition for the atmosphere (see table 3.1). Convection, which is clearly an important process in view of the extensive nature of the clouds in the atmosphere, is probably as strongly dependent upon the internal heat source resulting from the production of energy by gravitational contraction as upon absorbed solar radiation. The observed mean planetary temperature for Jupiter is approximately 128 K which using equation (2.1) and with an albedo about 0.6 (table 2.3) confirms the hypothesis of an internal energy source by indicating that the planet is emitting at least $2-2\frac{1}{2}$ times as

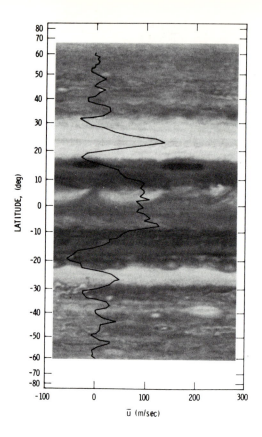

Figure 3.1. Profile of Jupiter's mean eastward winds, measured relative to the radio period of 9 hours 55.5 minutes, overlaid on a cylindrical projection of Jupiter. On the original colour projection it is possible to note the association of belts (darker bands) with cyclonic shear (northwest–southeast tilt in the northern hemisphere, northeast–southwest tilt in the southern hemisphere) and of zones (lighter bands) with anticyclonic shear; whereas a similar profile of Saturn's mean eastward winds shows little correlation between the belt and zone albedo/cloud patterns and the zonal wind profile. (After Smith *et al* 1981), see also the frontispiece.

much energy as is absorbed from the Sun. The schematic vertical temperature structure of the atmosphere shown in Figure 2.11 may be compared with that established from the Voyager spacecraft shown in figure 3.2. The very high temperatures observed at depth in Jupiter's atmosphere are a

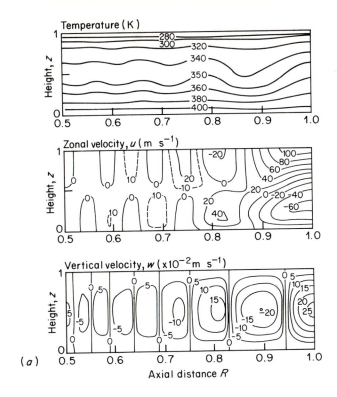

Figure 3.2. (*a*) Numerical simulation compared with (*b*) observations of the planet Jupiter. The contours of temperature, zonal wind and vertical velocity (*a*) show the large-scale, internally driven convection (after Chamberlain 1978). Compare also with figures 2.6, 2.11 and 3.3. The observed temperature profile in the upper layers of the atmosphere (*b*) indicates that the Great Red Spot is cooler than its surroundings. (Lower left): Examples of temperature profiles obtained by inversion of spectral radiances within the S(0) and S(1) hydrogen lines and the v_4 band of methane. The profile over the Great Red Spot was obtained from an average of six spectra; those at $+10°$ and $-15°$ latitude are from averages of approximately 60 spectra. (Lower right): Zonally averaged meridional temperature cross section of Jupiter. Data from two global maps taken respectively 2 days before and 2 days after encounter were combined to obtain this result. Both figures are from Voyager 1 (after Hanel *et al* 1979). (*c*) Colours in the belts and zones of both Jupiter (left) and Saturn (right) appear to be associated with cloud layers at specified depths in the atmospheres. The superimposed temperature profile shows that a temperature inversion marks the top of the convective region. The chemicals causing the coloration have yet to be identified (after Ingersoll 1981).

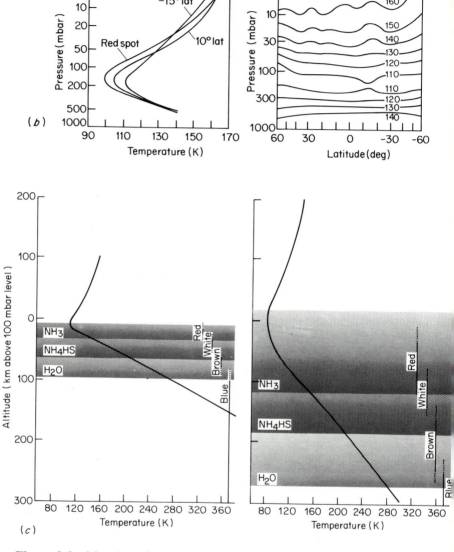

Figure 3.2. (*Continued*)

direct consequence of the high-energy flux from the planetary core. Smoluchowski (1967) has calculated that gravitational contraction of approximately 1 mm per year in Jupiter's radius would provide the required energy source. Similar internal heat sources may be of significance for the evolutionary history of all the jovian planets (Trafton 1981).

Table 3.1. Composition of Jupiter's atmosphere (after McElroy 1975), see original for references to sources.

Gas	Abundance[a]	Mixing ratio[b]	Wavelength[c]
H_2	67	1	0.82
He		0.2	0.06
CH_4	45	$\sim 10^{-3}$	1.1
NH_3	13	$\sim 10^{-4}$	1.1
CH_3D	2.6	$\sim 10^{-7}$	4.6
C_2H_6	10^{-4}	4×10^{-3d}	12
C_2H_2	2×10^{-6}	8×10^{-5d}	13
PH_3	e	e	
H_2O	f	f	5

[a] The abundance of H_2 is expressed in km atm, while the abundance of CH_3D is given in cm atm. Abundances for all other species are in m atm.
[b] Mixing ratio by volume with respect to H_2.
[c] Approximate wavelength (μm) at which the gas was detected.
[d] These abundances and ratios are uncertain.
[e] Observed by S T Ridway, but abundance not determined as yet (see McElroy 1975).
[f] Positive detection reported by Treffers et al (see McElroy 1975).

Pollack et al (1979) have considered the formation process of the jovian planets and their satellites. All the planets are believed to possess central cores composed of rocky and/or ice mixtures. As noted in chapter 1 (and e.g. Williams 1975) the satellite systems of Jupiter resemble somewhat the whole solar system in terms of the density and chemical gradients. Pollack et al (1979) believe that this radial differentiation may have resulted from the very much higher flux from the parent planet early in the evolution of the solar system. The nature of the decline in internal energy may be of considerable importance in understanding the histories of the atmospheres of these massive planets. For instance, whilst it is clear that there has been progressive contraction and cooling of the gaseous envelopes it remains uncertain exactly how the chemical nature of the core may affect both the heat flow and the overall chemical composition of the atmosphere.

Pollack et al (1977) and Pollack and Yung (1980) note two interesting features of the origin and subsequent evolution of these atmospheres. Firstly, they suggest that capture of planetesimals during the formation process may have been significant for the jovian planets. It is possible that significant chemical inhomogeneities in the cores of the giant planets could have been caused by the 'sweeping-up' of small debris. These authors also describe the major effect likely to cause a change in the overall constitution of these atmospheres. This is the removal of He from the gaseous state by its inclusion into metallic hydrogen when (and if) the temperature of the metallic hydrogen in the core falls below about 10^4 K. It is unlikely that Uranus and Neptune could support internal pressures and initial temperatures high enough to achieve a metallic hydrogen phase but theoretically both Jupiter and Saturn could achieve these conditions. The immiscibility of helium in a liquid hydrogen melt would remove it to the planetary core and thus, the authors suggest, permit detection of an atmospheric evolutionary process in a depleted level of atmospheric He. The data obtained for Saturn from Voyager 1 reveal a hydrogen:helium ratio 4% lower than the 9:1 level established for Jupiter. This observation is in agreement with the radiation data for the two planets.

The suggestion that there is a greater concentration of mass in the core of Saturn is in agreement with the degree of rotational flattening. On Jupiter the equatorial radius is 6.5% greater than the polar radius, while on Saturn the figure is 9.6%. Saturn has a much larger excess of emitted thermal flux compared with radiation received from the Sun. This extra energy is partially generated by the viscous friction between the descending helium droplets in the metallic hydrogen. It seems likely that the extra mass of Jupiter has, so far, held its central temperatures too high for the inclusion of helium into liquid hydrogen. Calculations now suggest that this process began on Saturn approximately 2×10^9 years ago and may just be beginning on Jupiter now (Smith *et al* 1981).

The addition of internal energy and absorbed solar radiation gives rise to a strongly convective regime in the upper atmosphere of the jovian planets. Observational evidence from both surface and satellites (figure 3.1) suggests that dynamic instabilities (chapter 2) may be of fundamental importance in balancing the energy transfer in a predominantly zonal flow situation. Gierasch *et al* (1979) have investigated a theoretical model of Jupiter's atmosphere consisting of two continuously stratified layers. They found that a basic similarity exists between the energy transfer by baroclinically unstable eddy circulations in the Earth's jet stream and instabilities developing in the jovian atmosphere (see figure 2.6). However, the dynamics of the two circulation systems differ considerably because of the differing lower boundary conditions (i.e. solid surface for the Earth compared with a deeper fluid layer on Jupiter). Energy transfer mechanisms between the two fluid layers in the model and the extent of convective activity within the upper (cloudy) layer are strongly dependent upon one another. The mechanisms of outward transfer of angular momentum and the effects of the differential rotation in the atmospheres of both Jupiter and Saturn have been discussed by Drobyshevskii (1979).

There is now known to be a very strong eastward-flowing equatorial jet and considerable turbulent interactions within the north and south tropical zones (see the frontispiece and figure 3.1). Numerical simulations of the large-scale convective processes within Jupiter's atmosphere have been described by Williams and Robinson (1973) and Gierasch et al (1979). Examples of the calculated values of temperature, zonal winds and vertical motions are shown in figure 3.2. There are a number of problems associated with these simulations. For instance, to permit simulation of motions at the observed scale, the required outward energy flow from the planet becomes very much too high. The almost monotonic zonal wind profile of Saturn resembles more closely the velocity profile of the photosphere of the Sun than that of Jupiter (shown in figure 3.1). It is suggested (Smith et al 1981) that the reason for the differences could lie in the very much greater importance of the internal heat source on Saturn compared with Jupiter (the internal energy of Saturn is more than 2.3 times the incident solar flux). The amplitude of the equatorial jet on Saturn suggests that vigorous eddies should operate within this part of the atmosphere. The images from Pioneer 11 and Voyager 1 show instead fewer small-scale circulatory features (cf the Great Red Spot and the white ovals on Jupiter). This paradox led to the suggestion that the atmosphere of Saturn may have a very much thicker haze layer than that of Jupiter or that the atmospheric colourants (chromophores) could be significantly less abundant in Saturn's atmosphere. However, analysis of Voyager 2 images indicates such a haze layer does not exist. The supposition is therefore that the chromophores are better mixed in the saturnian atmosphere. Colour enhanced images of Saturn do show a belt and zone structure similar to that on Jupiter and considerable preprocessing can reveal small-scale convective features which drift with the jet stream.

Two types of theoretical model have been applied to the atmospheres of Jupiter and Saturn. The stellar-type model seeks to interpret the observed atmospheric motions as the direct result of deep movements driven primarily by internal heat (figure 3.2(a)). Ingersoll (1981) has reviewed this and the alternative 'decoupled' theory. In the latter model an upper (thin) layer of these massive atmospheres is treated mathematically in a manner similar to that employed in meteorological models of the Earth. The upwelling heat is assumed to be independent of latitude and to have a negligible effect upon the atmospheric motions which are all driven by the latitudinal variation of incident solar radiation. Both models reproduce the banded structure but the 'decoupled' model fails to generate large persistent eddies analogous to the Great Red Spot and white ovals. Ingersoll (1981) suggests that the considerable difference between the internal structure and energy release of Saturn and Jupiter may soon make discrimination between these two model approaches possible. Both Jupiter and Saturn exhibit a north/south hemispheric asymmetry.

Complex chemical and photochemical reactions are believed to occur at many levels in the jovian atmospheres. The processes involving CH_4 and NH_3 are of particular interest since both these gases are relatively abundant; both exhibit spectral features which are easily observable from the Earth and from spacecraft and both have formed major components in the gaseous mixtures in the Urey–Miller-type (chapter 4) high-energy uv and electrical discharge experiments (e.g. Schwartz 1981). The products of this type of process no doubt contribute significantly to the exotic colours seen in the jovian atmospheres. There may also be longer-term chemical modification as a result of precipitation of more complex molecules into deeper atmospheric layers. However in these layers they will be destroyed by heat, and recycling could occur. A change which could be considered 'evolutionary' would therefore be the result of the interplay between photochemical production and dynamic mixing processes.

The atmosphere of Saturn presents a less colourful image than that of Jupiter (figure 3.2) but it also exhibits the banded structure presumably resulting from similar, but possibly weaker, convective activity. Gaseous methane is one of the most important secondary components of this atmosphere which is, like Jupiter's, composed primarily of hydrogen. The emission features of CH_4 and many of the photochemical derivatives (e.g. C_2H_2 and C_2H_6) are apparent in the infrared spectra observed by Voyager 1 (Hanel *et al* 1981). The molecular constituents identified in the saturnian atmosphere are listed in table 3.2. Although ammonia has been positively identified, its spectral features which dominate the jovian infrared spectrum are very much weaker in the case of Saturn. It is suggested that this is the result of condensation of ammonia in the upper troposphere, itself a direct result of the lower temperatures in the atmosphere of Saturn. Figure 3.3 illustrates the temperature lapse rates for Jupiter and Saturn as established by the Voyager spacecraft and

Table 3.2. Atmospheric composition of Saturn from Voyager 1 (after Hanel *et al* 1981).

Gas	Band	Wavenumber (cm^{-1})	Approximate mole fraction
Positively identified			
Hydrogen (H_2)	S_0, S_1	300–700	0.94
Helium (He)		200–600	0.06
Ammonia (NH_3)	Rotational	~200	2×10^{-4}†
Phosphine (PH_3)	ν_2	990	1×10^{-6}
Methane (CH_4)	ν_4	1304	8×10^{-4}†
Ethane (C_2H_6)	ν_9	821	5×10^{-6}
Acetylene (C_2H_2)	ν_5	729	2×10^{-8}
Tentatively identified			
Methylacetylene (C_3H_4)	ν_9	633	
Propane (C_3H_8)	ν_{26}	748	

† Assumed value.

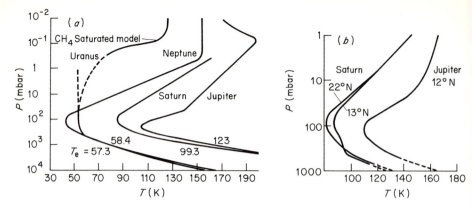

Figure 3.3. Temperature profiles for the jovian planets (*a*) as calculated from Gautier and Courtin (1979) and (*b*) by the infrared probe of Voyager 1, the broken curves are extrapolations along adiabats (after Hanel *et al* 1981). The predicted curve in figure 2.11 compares favourably with the observations.

compares these with retrieved lapse rates for the other jovian planets.

Despite the fact that large-scale changes in the atmospheric mass and composition of the jovian planetary atmospheres are unlikely to have occurred over the lifetime of the solar system, it is important to recognise the smaller-scale changes which are continuously taking place. For instance the complexity of the organic compounds in the atmospheres of both Jupiter and Saturn points to photochemical reactions in the stratospheres of greater significance than similar processes in the stratospheres of the smaller planets where oxygen is present. The condensation of NH_3 which is lower in the atmospheres of Saturn than Jupiter underlines the importance of the knowledge of temperature and density profiles with height for successful interpretation of current and past atmospheric processes. There is even a suggestion that a type of 'pseudo-surface'–atmosphere interaction may be significant for Saturn (Hanel *et al* 1981). Under the assumption of negligible heavy elements the Voyager 1 preliminary results indicate that the mass fraction of He in Saturn's atmosphere is considerably lower than that in Jupiter's atmosphere (0.11 cf 0.19) indicating a significant gravitational separation of helium and hydrogen inside Saturn as described above.

The interaction between the solar wind and the upper atmospheres of both Jupiter and Saturn is severely constrained by the existence of planetary magnetic fields as demonstrated by the aurorae seen in the Voyager images of Jupiter. However, it has recently been suggested (Hanel *et al* 1979) that the observations of germane, GeH_4 may indicate direct interaction with the solar wind. Strobel and Yung (1979) have suggested that the Galilean satellites may be a source of the CO detected in the upper atmosphere of Jupiter. Also, as has been described in chapter 2, the interactions between their upper atmospheres, the magnetic field lines and certain of their satellites can lead to both modifications to the atmosphere and also to recycling of the gaseous constituents of the satellite (see §2.1.5 and figure 2.8).

3.1.2. *Uranus and Neptune*

All estimates of the atmospheric masses and constituents of the remaining jovian planets (Uranus and Neptune) have been made from surface-based observation of spectra primarily in the infrared region (see table 1.4 and figure 3.3). Fink and Larson (1979) review the most likely atmospheric configurations for these two planets and compare their results with the available data. They suggest that there are considerable differences between these two atmospheres even though the spectra of both are dominated by CH_4 bands. Trafton (1981) reviews recent observations of gaseous constituents. Hydrogen, methane and probably helium are the major gases. Ammonia seems to be underabundant in the Uranus atmosphere by two orders of magnitude compared with the solar levels. It is likely that the upper atmosphere of Uranus is optically thin and may be relatively depleted in CH_4 whereas the atmosphere of Neptune has a cloud layer above which exists a significant amount of molecular hydrogen (Macy 1979, Pilcher *et al* 1979). Additionally it has been suggested (Tokunaga *et al* 1975) that Neptune's atmosphere exhibits a feature probably associated with emission from C_2H_6.

Microwave observations of Uranus led Gulkis *et al* (1978) to suggest that NH_3 is depleted as a result of reactions causing the precipitation of solid NH$_4$HS. However Fink and Larson (1979) report that both NH$_3$ and H$_2$S are depleted suggesting a real deficiency of nitrogen. Ground based spectral measurements for Uranus suggest that its internal energy source is very much less than that of Jupiter, Saturn and Neptune (Trafton 1981). Gautier and Courtin (1979) have constructed 'retrieved' temperature profiles for the four jovian planets (figure 3.3) and these compare favourably with the more recent satellite (Voyagers 1 and 2) estimates of the temperature variation with height. Infrared spectra for Uranus do not contain the expected methane emission feature at 7.7 μm (see §4.2.7) and it is therefore deduced that the stratosphere of Uranus is very much colder than that of the other jovian planets. The peculiar orientation of Uranus (its axis is inclined at 97°55′) could give rise to very large seasonal changes. Gautier and Courtin (1979) suggest that Uranus radiates only 1.3 times the solar energy it absorbs compared to values of 1.8, 2.3–3.0 and 2.6 for Jupiter, Saturn and Neptune. This considerable range could originate from differences in the chemical and physical composition of the central cores of the jovian planets (Podolak and Cameron 1974). It is not obvious that the chemical processes which control evolution on these major planets will be the same. It is also possible that the history of Uranus has been severely modified by its unusual orientation and by the lack of an internal heat source comparable with the energy absorbed from the Sun. It might be expected, therefore, that the chemical composition of the atmosphere of Uranus will differ considerably from those of Jupiter and Saturn.

3.2. A Numerical Model for Atmospheric Evolution

On the terrestrial planets and some moons the average planetary surface temperature, T_s, can be used as a measure of the prevailing climatic environment. The Earth is the only planet for which we have quantifiable and spatially varied geological data. Even for our own planet there is a dearth of data and the best method of interpretation is not always clear. Rocks are spatially and temporally separate, inhomogeneous and possibly unrepresentative of global mean conditions. Using the very early Precambrian deposits as a record of average climatic conditions on the Earth poses particularly serious problems. For instance, Knauth and Epstein (1976) interpret certain chert deposits as evidence of elevated early surface temperatures whilst the presence of (metamorphosed) sedimentary material in the oldest Precambrian formation (Isua, Moorbath *et al* 1973, Schidlowski *et al* 1979) could alternatively be interpreted as evidence for ambient surface temperatures not dissimilar from the present day (Henderson-Sellers and Meadows 1978).

Data pertaining to the histories of the planetary atmospheres fall into two groups: volatile inventories and residual constituents, and evidence for surface temperature changes. Pollack and Black (1979) consider the former in some detail with particular reference to the data on Venus. However more recent measurements made by the Pioneer Venus spacecraft (Donahue *et al* 1981) suggest that noble gas ratios between Venus and the Earth differ considerably (see §3.4). Here the results of computer programs designed for the computation of surface temperatures for mean global conditions are considered and a comparison is made between these results and the available surface temperature data.

The surface temperature at any epoch can, in principle, be calculated in two stages. The first involves the computation of the effective temperature of the planet (T_e) from equation (2.1). All the factors are straightforward except for f, which is discussed below. For a rapidly rotating planet with a thick atmosphere, the area emitting radiation is taken to be $4\pi R^2$ (where R is the planetary radius). For a slowly rotating planet with a thin atmosphere, the corresponding emitting area is $2\pi R^2$. Since the area receiving solar radiation is πR^2, the flux factor (defined as the ratio of the emitting area to the absorbing area), f, is generally assumed to be either 4 or 2. But the choice need not always be so obvious. In physical terms, the value assigned to the flux factor depends on the ratio of two characteristic times. The first (τ_h) is the time required for a planetary atmosphere to radiate away a major part of its heat content; the second is the rotational period of the planet (τ_r). If $\tau_h/\tau_r \approx 1$, neither $f = 4$, nor $f = 2$, represent good approximations and it becomes necessary to use a transitional value (for a more detailed discussion of f see Henderson-Sellers (1976) and §2.1.2).

The second stage of the calculation relates T_e to the surface temperature, T_s, by equation (2.4). The computer programs that solve equations (2.1) and (2.4) for T_s over a wide range of the relevant input parameters and at any epoch apply a method similar to that of Sagan and Mullen (1972) but include a re-analysis of the laboratory data available on infrared absorption as a function of pressure (for further details see Henderson-Sellers 1976, Henderson-Sellers and Meadows 1976).

The average surface temperature of the planet, T_s, is obtained, allowing for the 'greenhouse' effect, by dividing the emergent flux into two parts: one emitted from the surface through 'windows' in the infrared absorption spectrum and one emitted from the top of the atmosphere at a temperature of T_1 (from Eddington's approximation, equation (2.12)). Thus we have

$$\frac{S}{f}(1-A) = \sum_{\lambda_i} eB_{\lambda_i}(T_s)\,\Delta\lambda_i + \sum_{\lambda_i} B_{\lambda_i}(T_1)\,\Delta\lambda_j \qquad (3.1)$$

where the infrared absorption spectrum is approximated by a step function, so that there is total absorption at the wavelength λ_j and zero absorption at the wavelength λ_i.

Here a method for the calculation of average surface temperatures for all the terrestrial planets at any stage during the history (and into the future) of the solar system is described. The surface temperature is a function of five parameters, four of which are interdependent, and are also functions of the value of T_s. These parameters are: the solar flux (incident at the planetary distance), S; the spherical albedo, A; the flux factor, f; the surface infrared emissivity, e; and the atmosphere (its total pressure, chemical composition and the partial pressures of its constituents). The latter four are related by various feedback loops to the surface temperature. The solar flux does not depend upon T_s, but is, of course, different for different planets and at different epochs in their development. A number of common-sense measures can be used to limit the combinations of parameters possible. For instance, it seems unlikely that a planet could ever take a very high albedo together with a low value of flux factor. This is because a high value of A (say above 40%) implies a cloudy atmosphere which is likely to mean that the 'blanket effect' of the atmosphere is efficient enough to allow the night-side radiation to be non-negligible. The reverse (high f and low A) cannot be ruled out, since a rapidly rotating terrestrial planet with a tenuous atmosphere will be in this category. The surface infrared emissivity depends upon what is likely to condense on to the surface which, in turn, depends upon the atmospheric constituents and the surface temperature (see chapter 2).

One method of considering the evolution of T_s is to construct a time-dependent curve for individual planets (as is described in §3.4). This approach cannot include variations in the physical properties of the planets or in solar evolution. An alternative method is presented in the appendix. Furthermore this method is independent of particular degassing and geochemical models (cf Hart 1978). Isopleths of surface temperature have been calculated and drawn using results from the model. The axes chosen are the incident flux at the planet (thus allowing variations in time, planetary distance and stellar model) and the absorber amount in the atmosphere. The latter is somewhat arbitrary, but an attempt has been made to cover most of the likely combinations of carbon dioxide and water vapour, since these are the most likely atmospheric constituents for the terrestrial planets. Zero absorber concentration will give the value of the effective temperature. The other discrete absorber amounts chosen are 10 mbar of CO_2, 100 mbar of CO_2 + 0.1 mbar of H_2O, 100 mbar of CO_2 + 1 mbar of H_2O and 100 mbar of CO_2 + 10 mbar of H_2O. If all the other parameters (A, f and e) are to be allowed to vary the multi-dimensional nature of the results also requires that a number of separate figures are drawn. Some parameters are more closely related than the others. In particular, groups of values of A, f and e have been considered. Table 3.3 lists the pairs of A and f values and gives the value of e taken in each case (a dash indicates that the pair of A and f values were assumed to be unlikely). The graphs of lines of equal surface temperature were plotted for

Table 3.3. Value of infrared emissivity, e, for various (likely) combinations of the flux factor, f, and albedo, A. See also figures A1–A5.

Albedo, A	Flux factor, f			
	2.0	2.5	3.0	4.0
0.07	0.90	0.90	—	0.90
0.17	0.93	0.93	0.93	0.93
0.30	—	—	0.95	0.95
0.70	—	—	—	0.95

all 10 combinations of A, f and e for incident fluxes ranging from zero to 4.0×10^3 W m^{-2} (approximately twice the flux incident at Venus now). These isopleths of temperature are displayed, in degrees K, in the appendix (figures A1–A5). No consideration of neutral broadening has been included. The effect of this has already been noted. The particular combinations of CO_2 + H_2O chosen will not apply to every planet, but the series of figures permits an interesting description of certain evolutionary tracks. The use of varying flux amount along the horizontal axis allows a discussion of either the time evolution of one model planet (the flux variation being caused by the evolution of the Sun), or of zones of surface temperatures throughout a planetary system (this time, the flux variation is caused by differences in planetary distances). A particular zone might correspond, for instance, to a planet possessing conditions suitable for the evolution of life. This will be discussed later.

The usefulness of this method of presentation of results is illustrated in the appendix by discussion of various planetary evolutionary tracks and also by the consideration of life-bearing conditions on these and other planets. One advantage of this method of displaying the results is that emphasis is given to planetary properties which are often either overlooked or incorporated in an *ad hoc* fashion for individual cases. For instance the treatment allows the effects of changing albedo and emissivity to be observed. It is extremely difficult to estimate the way in which planetary albedos evolve since they are functions of temperature lapse rate, surface temperature, meridional flux gradient, rotation rate, etc. Therefore it is important to be able to estimate the magnitude of the effect of possible albedo changes.

The range of combination of planetary properties displayed in the appendix (figures A1–A5) is extensive. However, it is important to note that while incorporation of certain feedbacks (e.g. albedo variations through cloud and surface changes) is fairly complete, other important effects are excluded partly because of lack of space. The most notable omissions are the interactions between the atmospheric constituents themselves. Photochemical reactions are not computed in the model. Account is not taken of the relationship between surface temperature and the partial pressure of the volatiles. This is because the links are difficult to quantify for CO_2 on the Earth (see e.g. Walker *et al* 1981) and impossible to generalise for all the planets. The selected levels of CO_2 are somewhat arbitrary but illustrative. The amount of water vapour in the atmosphere is a function of the surface and vertical temperatures. These will change dynamically. The present-day relative humidity profile for the Earth (i.e. approximately 50% saturation at the surface rising to 100% at the tropopause) could have been included as is done in many climate models but this was believed to be too particular a constraint. Typically planets with open water areas on their surface could be expected to possess between 1 and 10 mbar of atmospheric water vapour. Three of the model planets considered bear a significant relationship to the three terrestrial planets possessing atmospheres. The generality of this method of presentation is underlined but comparison can be made between these model evolutionary histories and those for Venus, the Earth and Mars (presented in §3.4). Each planetary evolution, denoted by a numbered symbol (a triangle, cross, circle or square) is described fully in the appendix.

The primary conclusion that must be drawn from the model evolutionary histories discussed in the appendix relates to the *type* of environmental stability achievable for the different planetary systems. In each case an equilibrium state (as typified by temperature) results, but only in the case of model 2 in which a major volatile (water) condenses on to

91

the surface does the temperature regime remain stable and hospitable.

The importance to the long-term evolutionary history of a particular planet–atmosphere system of any one of the physical characteristics is itself a function of the state of other variables. For instance, the model histories described in the appendix (especially models 2 and 3) assume a comparatively slow degassing and/or atmospheric accumulation (cf §2.1.2). In the case of a planet such as Mars the relative rate of atmospheric mass build-up and changing solar luminosity could be critical. It is for this reason that the two extreme histories of terrestrial planetary mass increase were described in §2.1. The model histories presented in the appendix are made more complex by the necessity to include changing values of S, the absorber amount, A, f and e in computations of T_s and also to consider the feedback effects between these parameters (see §2.3). There are two situations in which at least some of these intricacies are removed: (i) instantaneous or very rapid atmospheric evolution (§2.1.1) and (ii) when the planetary mass is too small to retain the gases evolved. In the latter situation only the solar flux will change with time. The planetary flux factor will remain at $f = 2$ and large surface temperature variations are to be anticipated. In §§3.3 and 3.4 these model results are applied to the terrestrial planets.

3.3. Tenuous and Transient Planetary Atmospheres

3.3.1. Mercury and the Moon

The Earth's moon and the planet Mercury have considerably smaller masses than Venus and the Earth (approximately 10^{23} kg cf 5×10^{24} kg, see table 1.1). Retention of gases around either of these bodies has already been shown to be unlikely in the discussion of escape velocities (§§2.1 and 2.2). The greater gravitational attractive force exerted by Mercury is opposed by the very high surface temperatures caused by its proximity to the Sun. Even in the earliest stages of planetary evolution the surface temperature on Mercury $T_s \simeq T_e \simeq 480$ K (i.e. so much higher than the planets described in §3.2 that it could not be plotted on figure A1).

There is, however, one mechanism of atmospheric enhancement associated with the planet–Sun distance: the contribution of elements directly from the solar wind. Although this is a very important element in the mass balance of the lunar atmosphere, Mercury, despite its position closer to the Sun, receives a smaller flux of material from this source. This is because the solar wind does not reach the surface of Mercury as it is deflected at an altitude of about 1500 km by the planet's magnetic field. (Proof of the existence of this field was one of the major results of the Mariner 10 spacecraft mission in 1974 although it had been predicted for a number of years before the flyby.) The input of protons from the solar wind in the case of Mercury forms a direct addition to the upper atmosphere.

One of the major gaseous constituents of the very tenuous atmospheres on both Mercury and the Moon is helium (see table 1.4). The helium concentration in the lunar atmosphere shows a very close correlation with solar activity and thus it is believed that the solar wind, in the case of the Moon unaffected by a magnetic field, is the prime source of He and also of Ne. Similarly chemical reactions between the solar wind and lunar surface grains (Hoffman and Hodges 1975) may produce very small amounts of, for example, CH_4, NH_3 and CO_2. The lunar rock and soil samples returned by the Apollo missions are dry and volatile poor. If the surface rocks can be taken as representative of the state of the interior this suggests that lunar volatiles escaped early in the lifetime of the solar system.

The interactions between the constituents of these tenuous atmospheres and the solar wind, often termed 'sweeping' is a major sink for atmospheric gases. For instance, Kumar (1976) suggests that the upper atmosphere of the Moon is controlled by ionisation followed by solar wind sweeping reactions whilst he believes that the circulation of elements by Mercury's magnetosphere could play a crucial role in the removal of gases from this planet.

Degassing rates and, indeed, the likely results of degassing on the Moon and Mercury must be considered in the light of the increased knowledge of their geological histories (e.g. Guest *et al* 1979). Despite considerable controversy about the nature and timing it is reasonable to expect any evolutionary model for these atmospheres to include an extensive bombardment/vaporisation and outgassing epoch early in their histories.

However, if the periods of traumatic bombardment and internal activity of either body coincided (e.g. figure 2.1(b)) it is possible that a significant atmosphere could have existed transiently. In the case of Mercury's history the point at which the magnetic field was 'born' is clearly of considerable importance. Pollack and Yung (1980) suggest that the Moon could have outgassed up to 0.1 m of H_2O (and, by analogy with Mars, approximately 1 mbar CO_2). The temperature evolution of the Moon, assuming that it was by this stage already in the Earth's orbit, must have started in the same way as that of the model planet 2 (see the appendix). Thus its initial surface temperature would be given by $T_e = T_s \simeq 300$ K. The prolonged existence of an atmosphere/hydrosphere only supported by a total volatile inventory of the order of 0.1 m of H_2O seems most unlikely. A fuller discussion of the operation of surface–atmosphere feedback effects in this type of very tenuous atmospheric configuration is also discussed in chapter 5. Although the predominant processes were probably different for Mercury a similarly short-lived evolutionary history seems most likely.

The present-day atmospheres of both these small bodies (see table 1.4) are believed to be in quasi-equilibrium with the production and removal processes which still exist. Kumar (1976) calculated that the present-day outgassing rate of CO_2 and H_2O is at least four orders of magnitude less than similar juvenile CO_2 and H_2O evolution on the Earth. The rate constants for current degassing are, however, extremely difficult to estimate (e.g. Walker 1978b). For both bodies the primary source and sink of all gases is now the solar wind. Any surface temperature plot for bodies with such thin atmospheres would be almost meaningless. Indeed the usefulness of the concept of an 'average global surface temperature' for either the Moon or Mercury is difficult to defend. It is possible to argue that these planets have probably existed in their present-day state for the majority of the lifetime of the solar system and, further, that any phase of atmospheric enhancement was much too short to be of significance for the surface environmental conditions or for the planetary evolution as a whole. It is, however, interesting to note that despite a general acknowledgment of the quiescent nature of the Moon there are continuing observations made of 'transient lunar phenomena': effects which appear to be associated with surface and/or atmospheric activity. Mills (1980) suggests that 'smoke' created by very weak outgassing leading to the lifting of surface lunar dust into the atmosphere may be a plausible mechanism for these phenomena.

3.3.2. *Pluto*

The presence of CH_4 frost on the surface of Pluto has been reported (Trafton 1981). Lebofsky *et al* (1979) have made detailed infrared photometric observations of Pluto. Their results appear to confirm the presence of CH_4 in some form but their spectrum of the planetary surface does not compare well with laboratory spectra for pure methane frost. Fink *et al* (1980) report the detection of a CH_4 atmosphere on Pluto. Considerable difficulties exist in establishing the size and density of the planet itself (Trafton 1981) which has led to diverse interpretations of some of the data. Golitsyn (1979) notes that a theoretical value of the surface pressure on Pluto can be made. He calculates that for equilibrium methane vapour pressure at the subsolar point (taken to have a temperature of 60 K) the surface pressure is approximately 10 Pa (=0.1 mbar). However, this value will be changed due to the large orbital variations of Pluto (perihelion occurs in 1989). Golitsyn (1979) suggests ranges of 50–63 K and 35–45 K for the subsolar and mean planetary temperatures, respectively. At the points of increased solar flux it might be anticipated that the atmospheric (methane) pressure will increase. The presence of other gases, such as neon has been suggested (Hart 1974, Golitsyn 1975) but more accurate estimates of planetary size make these calculations inappropriate. It is possible that the highly eccentric nature of Pluto's orbit may produce the major mechanism for atmospheric evolution. The increase in the solar flux incident at the planet may just act as a trigger for atmospheric build-up. The

Table 3.4. Characteristics of the 'icy' satellites. (Footnote after Smith *et al* 1979.)

Satellite	Distance from parent	Radius (km)	Density (kg m^{-3})	Geometric albedo	Surface features and/or atmospheric gases
Jupiter (total 13)					
Io[†]	5.90 (R_J)	1820 ± 15	3530	0.73	Active volcanoes[†]; craters and scarps, SO$_2$ (see footnote)
Europa	9.40 (R_J)	1565 ± 15	3030	0.68	Frozen water ice (? 100 km); surface smooth
Ganymede	14.99 (R_J)	2640 ± 15	1930	0.34	High albedo 'ray' features? H$_2$O ice, tectonic crustal, creep and cratering
Callisto	26.33 (R_J)	2420 ± 15	1790	0.13	Heavily cratered, large features similar to lunar mare
Saturn (total 15)					
Mimas	3.08 (R_S)	195 ± 5	1200 ± 100	0.6 ± 0.1	Craters and shock trough features (one very large crater)
Enceladus	3.97 (R_S)	250 ± 10	1100 ± 600	1.0 ± 0.1	Few craters—(?) glacial feature cutting craters—? active surface
Tethys	4.91 (R_S)	525 ± 10	1000 ± 100	0.8 ± 0.1	Many craters >200 km and long trench $\frac{3}{4}$ way round moon
Dione	6.29 (R_S)	560 ± 10	1400 ± 100	0.6 ± 0.1 ⎫	Craters and 'whispy' albedo features
Rhea	8.78 (R_S)	765 ± 10	1300 ± 100	0.6 ± 0.1 ⎭	
Titan	20.40 (R_S)	2560 ± 26	1900 ± 100	—	Haze and clouds; N$_2$, CH$_4$, ? Ar (see table 3.5)
Hyperion	24.70 (R_S)	145 ± 20	—	0.3 ± 0.1	Irregular shape—long axis points to Sun
Iapetus	59.30 (R_S)	720 ± 20	1200 ± 500	0.5 ± 0.3	Albedo highly asymmetric, brightside cratered, ? painted by material from Phoebe
Neptune (total 2)					
Triton	15.90 (R_N)	2000	4178	0.2–0.4 (?)	? Albedo markings vary CH$_4$
Pluto (total 1)					
Charon	(variable)	1300	1200	0.4	? CH$_4$ at perihelion

† Io volcanic plume inventory from Voyager 1. (See frontispiece to this chapter.)

Plume	Latitude	Longitude	Approximate height‡ (km)	Remarks
1	−28°	248°	270	Symmetric, first discovered
2	10°	300°	100	Asymmetric, diffuse
3	−5°	145°	100	Symmetric
4	20°	168°	100	Asymmetric, diffuse
5	27°	97°	100	Diffuse
6	16°	109°	100	Symmetric
7	−33°	212°	100	Diffuse

‡ Heights are measured in visible light. In ultraviolet, images show a larger faint envelope rising to about twice this height (after Smith *et al* 1979).

freeze–thaw process and escape may produce long-term changes in the total volatile inventory.

3.3.3. *The Satellites of Jupiter and Saturn*

The term 'icy' satellites was coined to describe most of the major satellites of Jupiter and Saturn. These satellites are composed of ice–rock conglomerates but may have volatiles in tenuous atmospheres and condensed on to their surfaces. Although the term 'icy' is not strictly applicable to two of Jupiter's satellites, Io and Europa, which are known to be rocky, they are included here. Discussion of the satellites Titan and Triton will be found in §3.4. The satellites to be considered are therefore the Galilean satellites of Jupiter, the six largest satellites of Saturn, excluding Titan, and the recently discovered satellite of Pluto, Charon (Christy and Harrington 1978). The properties of these satellites are listed in table 3.4. With the exception of Io, none of these satellites are believed to possess atmospheres (Parmentier and Head 1979, Cruikshank 1979). For example, a stellar occultation by Ganymede which was observed with the extreme-UV spectrometer during the Voyager 1 encounter with Jupiter indicated that this satellite's atmosphere is, at most, an exosphere (Broadfoot *et al* 1979). The radius of Charon is calculated to be approximately 1300 km and its albedo about 0.4. It has been estimated from the work of Consolmagno and Lewis (1976) that the density of Charon is likely to be about 1.2×10^3 kg m^{-3}. Thus its small size and low gravity combined with the very low temperatures at the orbital distance of Pluto make the existence of an atmosphere around this body most unlikely (Golitsyn 1979).

The Voyager 1 and 2 images of the Galilean satellite Io showed (see the frontispiece of this chapter) a rocky surface clearly undergoing intensive tectonic and volcanic activity (see table 3.4). The density of Io (3.53×10^3 kg m^{-3}) suggests a silicate planet, and spectral analysis of the volcanic plumes indicates that the major volatile being degassed is sulphur in the form of SO_2 (Morabito *et al* 1979, Fanale *et al* 1979). Slobodkin *et al* (1980, 1981) have successfully matched the spectral observations from Voyager 1 with spectra of SO_2

frost. More results of their work enable them to suggest that the freezing of these atmospheric aerosols took place at temperatures below 110 K. Numerical analysis of the particulate sizes and dynamics of the volcanic plumes (Wilson 1980) indicates that the activity is so intensive that a crustal layer of about 500 m deep is being reprocessed approximately every 10^5 years.

Jupiter is one of the brightest objects in the radio astronomer's sky. Fluctuations in the radio signals from Jupiter have long been known to be associated with the satellite Io. The jovian magnetic field has also permitted the formation and persistence of an extensive cloud of hydrogen atoms in partial toroidal configuration around Jupiter at the orbital distance of Io (see §2.1.5 and figure 2.8). As well as this plasma torus, there exist clouds of neutral atoms in the orbital region of Io. The major constituent is sodium but H, Ca and K have also been detected (Trafton 1981). Atmospheric Na has been observed on Io and estimates have been made of the surface density of SO_2 of the order of 5×10^{12} cm^{-3} at the subsolar point and at 2×10^{11} cm^{-3} or less globally (Trafton 1981). Pilcher (1979) has suggested that '"gardening" of Io's surface by meteoroid impacts is capable of exposing previously buried material to sputtering radiation and these may be a factor in the rate of water loss from the satellite'. It has also been suggested that the escape of water vapour from Io must be a direct result of the energetic volcanic activity. If Io was formed with the same fractional abundance of water as the Earth (a gross assumption) then it will have released about 100 m H_2O (Pollack and Yung 1980). There is little likelihood of incomplete degassing on Io because of the planet-wide tectonic activity. (Pearl *et al* (1979) have calculated that the whole planet is molten except for a very thin crust approximately 20 km thick.) The very energetic volcanic plumes produced (altitudes attained $\simeq$ 300 km, table 3.4) may lift any released H_2O vapour high enough to make Jeans escape a substantial and comparatively rapid loss mechanism. Despite this continuing outgassing there seems to be little evidence of accumulation of, for instance, N_2 or ^{40}Ar (Pollack and Yung 1980), although segregation and removal may also be affected by magnetospheric effects. Io's features may be summarised as follows:

the surface is continually renewed;

SO_2, Na, K identified and no N, C, H;

$75 \text{ K} \lesssim T_s \lesssim 130 \text{ K}$;

active regions have temperatures $\gtrsim 650 \text{ K}$;

tidal flexing is the most probable source of internal heat.

The other Galilean satellites and the larger satellites of Saturn possess, at most, tenuous atmospheres and transient activity (see table 3.4). The major features of the remaining satellites may be listed as:

surface not primordial;

surfaces composed of water ice overlying denser material;

T_s for Jupiter's satellites 80–130 K;

T_s for Saturn's satellites 55–100 K;

Enceladus and Europa may both have active surfaces.

Impact cratering and tectonic activity (either in the form of plate-type movement or volcanism) could be continuously producing very thin atmospheres similar (in density though not in chemical constituents) to that of the Moon. These satellites also illustrate the history of bombardment activity in the solar system. External and internal formation processes seem to occur together only on Ganymede (Parmentier and Head 1979, Guest 1980).

The icy surface of Callisto has been severely altered by impacts but the complete absence of volcanic activity contrasts strongly with the surface features on Io. The 'jigsaw' appearance of Ganymede suggests early or suppressed plate movements in which the darker areas are moving apart and the lighter material is filling by mantle upwelling. The lighter areas are less heavily cratered and therefore younger. Europa seems to have a water ice surface (Cassen *et al* 1979) which has been cracked around the whole planet into linear features. Few craters were seen in the Voyager images but this, it is suggested, could be because the surface does not retain impact features. The shattered appearance of this satellite could be a result of gravitational 'flexing' by the parent planet, Jupiter (see e.g. Ransford *et al* 1981).

The icy satellites of Saturn also exhibit crater and indeed tectonic features (Smith *et al* 1981). Most spectacular is the enormous crater on Mimas with a diameter about $\frac{1}{3}$ of the diameter of Mimas itself. Its surface is also covered by linear grooves which may have resulted from tidal interaction with Saturn. Its albedo suggests a surface covering of frozen water ice (table 3.4). Another unexplained contrast is seen between these features and the smoother and brighter appearance of the satellite Enceladus which is both adjacent to Mimas and closest to it in size. Voyager 2 images of Enceladus show a feature which is similar to glacial flow/trench erosion. The albedo of Iapetus shows a large hemispheric asymmetry ($0.03 \rightarrow 0.40$–0.50). This is believed to be the result of deposition of material from Phoebe favouring one side only. Smith *et al* (1981) conclude from similar analyses of the larger saturnian satellites that crater counts are consistent with the working model of a very large bombardment of all bodies in the solar system around 4.0×10^9 years ago (cf figure 2.1). This bombardment which ceased relatively quickly is responsible for the population I craters on these satellites and on the Moon, Mercury, etc. The densities of the icy satellites (table 3.4) may also be a function of possibly increased luminosities of the protoplanets of Jupiter and Saturn. Further examination of the accretion and condensation processes could possibly lead to a fuller understanding of the chemical state and tectonic activity on the terrestrial planets and on Titan.

3.4. Atmospheres of the Terrestrial Planets

The three largest terrestrial planets, together with Titan and possibly Triton, Neptune's larger satellite, compose a rich and diverse set of individual solutions to the problem of evolution of terrestrial-type planetary atmospheres. It is not certain whether the states viewed today are stable and now quasi-quiescent or whether today's view of each planet is simply a single 'snapshot' in a unique evolutionary history. This problem of the uniqueness of the planetary environments as observed is important to Man now because it relates to two questions which may be crucial for survival: (i) the likelihood of the origin of life and its subsequent evolution elsewhere and (ii) the stability of the climate to perturbations. The evolution of life is itself a serious perturbing influence in a planetary system as described in chapter 4 (see especially figure 4.4 which shows the major cycles of oxygen on the Earth and the dominating influence of the biota upon these cycles). The question of the necessary and sufficient conditions for life are considered in chapters 4 and 6 whilst the other shorter-term perturbing influences on planetary atmospheres are discussed in chapter 5. Here examination is made of the long-term evolutionary trends on the Earth, Venus and Mars. An attempt is made to try to establish how and why the planetary histories have differed and brief consideration of the possible evolutionary histories of Titan and Triton is undertaken.

3.4.1. Venus, the Earth and Mars

The primary factors likely to cause the differences observable in the present-day states of the terrestrial planetary atmospheres are: (i) differences in the mass, chemical composition and differentiation and tectonic state (see e.g. Head and Solomon 1981) of the planets themselves (see chapter 1) and (ii) different planet–Sun distances. The secondary or feedback effects resulting from these two major sets of characteristics may also be of significance (see chapter 5). The detailed consideration of the likely evolutionary sequence on the Earth, Venus and Mars leads to the conclusion that another factor—the precise level of incident solar radiation received by the planet at the time of and immediately after its formation—may also be critical for the subsequent evolution. It seems reasonable to assume (following the argument in §1.1) that the overall chemical compositions of Venus and the Earth are very similar and whilst Mars may differ it too is generally believed to have been formed from similar materials to the other terrestrial planets. Indeed most proposed volatile inventories for the planet Mars depend upon this assumption since the total volatiles available are generally calculated by scaling known or estimated ratios for the Earth or Venus (see discussion below of table 3.5).

Table 1.5 lists the volatile inventories for Venus and the Earth (present-day and an 'unweathered and abiological Earth') (after Owen 1978). The correlation between the latter and the Venus inventory is superficially good. However, it is as important to analyse the differences between these two planetary atmospheres as the similarities. For instance the Earth's atmosphere contains 0.78 bar of N_2 and the sediments contain roughly $\frac{1}{3}$ of this amount (according to Walker 1977) for a total of 1.04 bar of N_2. Venus' atmosphere is about 3.5% N_2 for a total of 3.2 bar (i.e. three times that of Earth). Earth's reservoir of carbonate rocks represents 5×10^{21} moles of CO_2 (Walker 1977) or about 45 bar of CO_2, compared to 90 bar in the cytherean atmosphere. These figures probably agree to within the error bounds on the carbonate reservoir.

Meanwhile, Earth has the equivalent of 300 bar of water in its oceans, compared with an upper limit of around 0.1 bar of H_2O on Venus. The agreement here is clearly not compelling. Volatile inventories for Mars have been estimated by a number of authors. Although the values differ, there is a general consensus that $H_2O : CO_2 : Ar$ ratios are comparable on the Earth and Mars (e.g. Walker 1977) and, indeed, among the C1 carbonaceous chondritic meteorites. Any estimates of total volatiles must be made with reference to the noble gas data. These data for the terrestrial planets are reviewed below. Following either a rapid evolutionary history (see §2.1.1) or a slower history including internal feedbacks (see §2.1.2), for both the Earth and Venus, can result in temperature histories which compare favourably with existing data (Henderson-Sellers et al 1980). In each case careful consideration must be made of the likely phase state and removal of each of the constituents. Figure 3.4, for instance, has to be considered in conjunction with the major cycles of other elements, especially oxygen on the Earth (chapter 4) and also the required surface conditions for the chemical and physical reactions involved. The lack of knowledge and understanding even about these pathways and rates for the present-day Earth has been underlined by the recent controversy regarding the fate of anthropogenically produced CO_2 (see chapter 2, figure 2.13, and Bolin et al 1979). The process of equilibration of atmospheric CO_2 levels (table 2.5) is controlled on the Earth by the ambient surface temperature and the presence of large ocean surfaces.

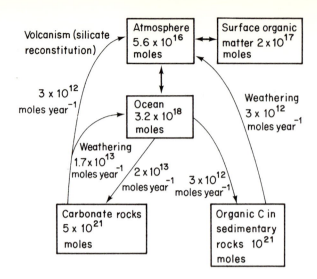

Figure 3.4. The major reservoirs (rectangular boxes) and cycling processes of carbon in the Earth's atmosphere–hydrosphere–biosphere system. Compare this with the effect of biospheric activity upon atmospheric CO_2 levels illustrated in figure 2.13 and also with table 2.5 (after Walker 1977).

Approximately 85% of the sedimentary carbon in the Earth's crust appears to be in the form of carbonate rocks. Without the hydrospheric processes producing e.g. limestones the atmospheric mass of CO_2 (from table 1.5) would be about 70 atmospheres. This assertion should not suggest that equilibrium does not occur on Venus. On the contrary it is believed that a similar quasi-stationary state exists on Venus with reactions directly between the surface rocks and the atmospheric CO_2. At an ambient surface temperature of about 700 K solid surface–atmosphere reactions such as:

$$\underset{\text{Anorthite}}{CaAl_2Si_2O_8} + CO_2 + 2H_2O \rightarrow \underset{\text{Calcite}}{CaCO_3} + \underset{\text{Kaolinite}}{Al_2Si_2O_5(OH)_4}$$

(3.2)

are much more rapid than in terrestrial conditions. Sagan (1975b) suggested that sintering (i.e. the welding together of surface grains with glass) of surface grains could be widespread on Venus. Mariner 5 and more recently Pioneer satellite observations of the Venus atmosphere suggested a dual exosphere and a flux escape rate of H from the planet not dissimilar from the rate for the Earth. If the amount of H_2O presumably originally degassed by Venus (~ 3 km by analogy with the Earth, table 1.5) has escaped after photodissociation it is likely that early in the evolution of Venus the mechanisms controlling H flux to and through the upper atmosphere were different (e.g. Chamberlain 1978). It is generally assumed (Pollack 1971, Walker 1977) that during the first 10^9 years of Venus' evolution surface temperatures (see §3.2 and the appendix) were held high enough (i.e. above 350 K) by other planetary constraints to force most of the volatiles into the juvenile atmosphere and thus, via a series of positive feedback effects this enhanced value of T_s is retained and even increased.

Noble gas data has already been referred to in support of the consensus view that the atmospheres of all the terrestrial planets are secondary (table 2.1(b)). The most recent data from Voyager 1 (see below) also suggests that the atmosphere of Titan is predominantly the result of planetary outgassing rather than capture during formation. Despite the general similarity in the data there are noticeable, and possibly very significant, differences observed on Mars, the Earth and Venus. Table 3.5 (after McElroy and Prather 1981) lists these abundances as derived from the Pioneer, Mariner, Venera and Viking spacecrafts. More recent data for the planet Venus have been derived by Donahue et al (1981) from Pioneer Venus. Figure 3.5 compares the abundances of rare gas isotopes for the three terrestrial planets with typical values for the C3V meteorites†. The large excess of Ne and ^{36}Ar which these authors observe in the cytherean atmosphere seems to be incompatible with simple models of terrestrial planetary outgassing from enriched surface veneers with a similar composition. The problem is particularly difficult in the case of neon, the ratios Venus:Earth being 3.5, ~ 30, 43 and 74 for ^{84}Kr, ^{132}Xe, ^{20}Ne and ^{36}Ar, respectively. The ratio of ^{36}Ar:^{84}Kr for Venus ($\simeq 1000$) more closely resembles the solar value ($\simeq 3710$) than the value for the Earth of approximately 48. These data seem to suggest that the noble gas

† The C3 meteorites are the least volatile-rich group of carbonaceous chondrites and have been implicated in planetary atmospheric formation by Anders and Owen (1977).

Table 3.5. Noble gas inventory and characteristics for terrestrial planets and potential external sources (after McElroy and Prather 1981). *Note* (see original for sources) [a] Total abundance assumes that 50% of ^{36}Ar remains trapped within the Earth; the efficiency of degassing for Venus (50%) is estimated using measurements of ^{40}Ar. [b] All mixing ratios are molar (vol vol^{-1}). Errors in the observations are not included. [c] Assumes ^{20}Ne/^{22}Ne = 8 and assumes a mean pressure of 7.5 mbar. The ratio N$_2$/^{36}Ar in the present atmosphere is approximately 5×10^3; the ratio given here refers to the original system and includes estimates for escape of N.

	Atmospheric ^{40}Ar (g g^{-1})	Atmospheric ^{36}Ar (g g^{-1})	Total ^{36}Ar (g g^{-1})[a]	^{20}Ne/^{36}Ar[b]	^{84}Kr/^{36}Ar	^{132}Xe/^{36}Ar	N$_2$/^{36}Ar	^{20}Ne/^{22}Ne	^{40}Ar/^{36}Ar	^{129}Xe/^{132}Xe
Venus	2.8×10^{-9}	2.4×10^{-9}	5×10^{-9}		4×10^{-4}	?	1.2×10^3	14	1	?
Earth	1.2×10^{-8}	3.5×10^{-11}	7×10^{-11}	0.52	0.021	7.5×10^{-4}	2.5×10^4	9.8	296	0.98
Mars[c]	7.0×10^{-10}	2.0×10^{-13}	4×10^{-12}	0.38	0.032	3×10^{-3}	9×10^5	?	3000	2.5
Sun:			3×10^{-5}							
Solar wind				35	5×10^{-4}	6×10^{-5}	1.2×10^2	13.3	<1	1.05
Flares				32				7.7		
Lunar fines			8×10^{-7}							
Meteorites:										
C			2×10^{-10} –3×10^{-9}	0.27	1.2×10^{-2}	1×10^{-2}	2×10^6	8		1–3
C3V			5×10^{-10}				1.5×10^5			
C3O			2.5×10^{-9}				2.5×10^4			
H, L, LL			2×10^{-12} –5×10^{-11}	low	1×10^{-2}	1×10^{-2}	2×10^6			
E			1×10^{-10} –1×10^{-9}							
Gas rich			2×10^{-11} –2×10^{-9}	20	10^{-3}	low		12.5		
'Neon-E'								<0.1		
Cosmic rays								2.6		

inventory becomes more solar-like as the Sun–planet distance decreases. McElroy and Prather (1981) postulate that enhanced values on Venus could be the result of exposure of the particles in the planetary nebula to the solar wind. It is clear that ^{22}Ne and ^{36}Ar are low on Mars compared to the Earth. The distance of Mars from the Sun and the time taken for planetary formation could combine to produce the lowered values for this planet. The noble gas abundancies on the Earth and Mars follow the planetary component exhibited by certain meteorites, though the lowered levels of ^{40}Ar in the martian atmosphere could be the result of degassing inefficiency (say only 10% as efficient as degassing on the Earth). McElroy and Prather (1981) suggest that Venus' noble gases are derived from the solar wind following implantation prior to planetary formation, while the low concentration of noble gases on Mars is due to differentiation and escape of volatiles in the pre-planetary regions of Mars. If the explanation of these noble gas abundances proposed by McElroy and Prather is accepted the complete volatile inventories for the terrestrial planets must be reconsidered. Of especial interest will be the possible effects that more rapid formation of the planet Mars (cf Venus) would have upon the total available H_2O.

Alternatively Pollack and Black (1979) have constructed a model for the formation of the terrestrial planets in which the incorporation of rare gases into the protoplanets of Venus, the Earth and Mars is controlled by the solar nebula itself. The amount of rare gases is found to increase linearly with pressure in the nebula but to drop exponentially as a function of temperature. This model seems to be contradicted by the Ar:Kr ratios for Venus shown in figure 3.5. Their model assumes that the planets grew in such a manner that their relative proportions of N and C are constant. Degassing rates can then be established from the ratio of ^{40}Ar:^{36}Ar. This value is difficult to determine even for the Earth. For instance, Pollack and Yung (1980) note that both the ^{40}Ar:^{36}Ar ratio, and incidentally also the absolute amount of ^{36}Ar, varies considerably according to where it is measured. The current atmospheric ratio of ^{40}Ar:^{36}Ar $\simeq 300$ cf the deep mantle rock measurements of about 450 (Ringwood 1979,

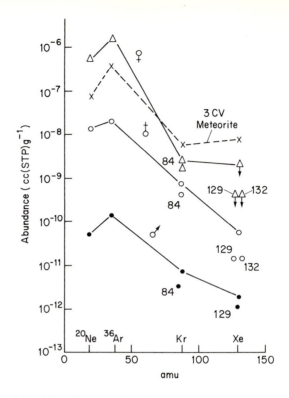

Figure 3.5. Abundances of noble gas isotopes on the terrestrial planets and from C3V meteorites (after Donahue *et al* 1981).

Stacey 1980) to ratios as high as 10 000 from upper mantle rocks. The total outgassing rates will scale roughly with the planetary bulk K content. Using data from Anders and Owen (1977) it therefore seems likely that comparable rates may be anticipated for Venus and the Earth with levels approximately $\frac{1}{4}$ of this for the Moon and possibly as low as $\frac{1}{20}$ for Mercury. At present no satisfactory model has been suggested that provides an explanation of both the total volatile inventories observed (e.g. C and N similar on Earth and Venus) and the noble gas data. Incorporation of two separate components of noble gases and volatiles in the accreting planets may be required to achieve agreement with observations.

One of the most complete analyses of the early evolution of the atmosphere of Venus is that by Pollack (1971). Subsequent models have dealt with the atmospheric chemistry in more detail (e.g. Walker 1977) but Pollack's paper is one of the few which attempts to consider interactive atmospheric and surface processes. His analysis has another advantage in that it uses a non-grey method of calculation of both the absorbed solar and the emitted infrared radiation. He also uses the absorption data of Howard *et al* (1956) and thus his results are directly comparable with those described in §3.2 and the appendix. Pollack (1971) considers only the infrared absorption due to water vapour in his model atmospheres which seems to be in conflict with more recent studies of the importance of the CO_2 contribution to the atmospheric greenhouse effect and his own assertion that certain of his computed temperatures would permit evaporation of all surface liquid water and thus remove a major sink mechanism for atmospheric CO_2 via the formation of sedimentary carbonate rocks. However, this consideration of only the effects of atmospheric water vapour has the advantage of permitting an analysis of such complex interactions as the relationships between albedo, relative humidity and the atmospheric lapse rate, which can be disguised when other effects are included. Despite the fact that some observational data acquired since these calculations were made are in conflict with Pollack's results (e.g. his calculation of the Earth's albedo as 0.39 cf table 2.3) the model is worthy of detailed consideration.

Atmospheric evolution is clearly a strong function of the rate and volume of outgassing. Pollack (1971) suggests that Venus outgassed comparatively quickly. Pioneer Venus observations of ^{40}Ar seem to support this proposition as do current interpretations of the observed hydrogen/deuterium ratio. Pollack's (1971) calculations indicate that the rapidity with which a runaway greenhouse effect could have occurred may have been dependent upon the percentage cloud cover established at the start of Venus' atmospheric evolution. Assuming an incident solar luminosity 30% lower than the present-day value, he suggests that runaway conditions are difficult to achieve for a situation of 50% cloud cover and, if the cloud cover were to rise rapidly to the present-day level of 100%, runaway conditions cannot be achieved even with the present-day input of solar radiation. Pollack (1971) believes that these calculations indicate a strong likelihood of more equable temperatures on Venus soon after planetary formation which may have persisted for a considerable geological period. Only identification of surface features which are either strongly indicative of at least one liquid water epoch (as seems to be indicated on Mars) or features which are both ancient and untouched by the type of erosion associated with a hydrosphere can finally establish the veracity of these calculations.

A synthesis of all the available topographical information obtained from radar mapping suggests that all the processes occurring on Earth *except erosion by rainwater* seem to be evident on Venus (Pettengill *et al* 1980). The surface is flatter† than had been expected with less than 40% of the total area exhibiting topographic features with a height range greater than one kilometre (figure 3.6). Elevated regions are attributable to both volcanic activity (e.g. Maxwell is a 'mountain' 11 km high) and to the differentiation of lighter 'continental' material (e.g. the regions named Ishtar and Aphrodite). A further interesting, but as yet inexplicable, feature of all the elevated areas is that at radio wavelengths they are considerably rougher than the low-lying areas. There is also substantial evidence for impact features remaining on the surface‡. For instance, even in the chaotic terrain on Aphrodite there is a feature which is almost certainly the remains of a huge impact event. The 'crater' is roughly circular and has a diameter of approximately 1800 km. Finally tectonic activity appears weaker than on the Earth since large-scale plate tectonics seem to be absent but there are a number of features which offer considerable evidence for tectonism. Some of the most compelling are the long, straight 'valleys' which closely resemble terrestrial rift valleys. Observations from Pioneer Venus permitted mapping of atmospheric lightning. This appears to be centred on the two elevated regions in Beta Regio (Rhea Mons and Theaia Mons) which may be volcanically active sites (Scarf *et al* 1980). Pettengill *et al* (1980) conclude that Venus 'has

† Pioneer Venus data indicate that the spherical harmonic coefficients are smaller for Venus than for the Earth.
‡ However recent radar observations suggest that some of the circular features around Maxwell are more likely to be collapsed volcanic caldera.

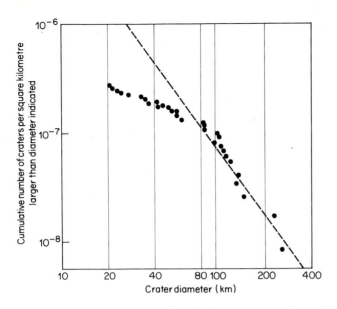

Figure 3.6. The cumulative distribution of ring-shaped features on Venus suggests that if they were created by meteorite impact, they conform to a theoretical model (broken line) for the number of impact craters that would have been made within the past 600 million to a billion years. The model is derived from counts of craters on the Moon, Mercury and Mars. The agreement is good for crater-like features that are larger than 80 km in diameter. Objects that would give rise to craters smaller than 20 km would tend to burn up in the dense atmosphere. The dearth of craters smaller than 80 km may be due to rock 'flow' (after Pettengill *et al* 1980).

evolved geologically as much as the Earth but not to the same degree'. The crater inventories made by the radar mapping teams seem to indicate that the surface of Venus is about 10^9 years old (figure 3.6). Hence, in the absence of (recent) water erosion, tectonic activity and surface–atmosphere 'weathering' must play a significant role in removing earlier surface features (McGill 1979). It is interesting to note that in their recent review Pollack and Yung (1980) reiterate the description of the temperature curves of Pollack (1971) for Venus. Pollack's results seem to be fairly sensitive to the total cloud cover which is an extremely difficult parameter to estimate even for slight perturbations to the Earth's current climate (see e.g. Roads 1978 and chapters 4 and 5). In the case of Venus the timing of the change in its orbital configuration and rotation rate may be an important factor controlling atmospheric dynamics and hence the planetary cloud cover (§§2.1.3, 3.2 and the appendix).

The time interval over which the breakdown and escape of the considerable quantity of water vapour initially believed to have been outgassed has been discussed in the literature (Ingersoll 1969, Walker 1977, Head and Solomon 1981). This time period may be as crucial in the history of the atmospheric evolution of Venus as are, for instance, the chemical changes subsequent to the production of free oxygen in the biosphere on the Earth (chapter 4). All descriptions of the 'runaway greenhouse' on Venus suffer from two problems: (i) achieving an adequate sink for the excess O_2 and (ii) explaining the final stages of H_2O destruction. These model histories have aspects which may also relate to the Earth. A major difficulty in constructing a satisfactory evolutionary history for the cytherean atmosphere is often explaining why the processes apparently *required* on Venus should not have occurred on the Earth. The models described in the appendix, in common with many such, rely upon the higher initial, and continuing, solar flux at Venus' orbit to 'explain' these discrepancies. This argument only seems reasonable if the divergence between the two planetary evolutionary histories occurred very early. This is because if, as has been suggested by Pollack (1971, 1979) the 'runaway greenhouse' did not occur until some time after planetary formation, then Venus must have passed through a stage of 'hospitable' Earth-like conditions. In itself this is an interesting hypothesis which has led Pollack (1971) and others to suggest that conditions could even have been hospitable enough to support life.

The occurrence of a fully developed atmosphere–hydrosphere system may even be useful in explaining the magnitude of the sinks for atmospheric oxygen which is believed to have resulted from photolysis of water vapour. For instance, Walker's (1975) model for the evolution of Venus can only operate if the weathering rate of O_2 from the cytherean atmosphere is equal to the current Earth rate. However, as will be described in detail in chapters 4 and 6, the Earth's atmosphere–hydrosphere system appears to be controlled by a number of

112

negative feedback processes. These operate in such a way that large excursions from surface conditions originally supporting the global hydrosphere seem to be opposed. It is very difficult to construct a model for Venus that allows passage from 'warm and wet' to 'hot and very dry' without implying that the Earth too is unstable, the latter being contrary to the geological data.

A more interesting sink for degassed water in the case of Venus has been tentatively proposed by geochemists (e.g. Holland 1962, Fricker and Reynolds 1968). This sink depends upon the fact that water is readily taken into solution in molten silica materials and that the efficiency of this removal process increases as a function of the water vapour partial pressure. The scheme for removal of water from the surface of Venus would then rely upon rapid and effective outgassing plus high insolation rates leading to early high partial pressures of H_2O vapour. An active lithospheric circulation must also be invoked so that the water taken into solution in the silica melts can be rapidly removed to deeper layers of the planet. However, the solubility of H_2O in silicate melts is only 5–6% so the efficiency of this process is questionable. The nature of the lithospheric circulation would presumably determine whether this is a final and irreversible sink or whether only oxygen is totally removed whilst the hydrogen is recycled to the surface. Since 'lubrication' by water of early tectonic activity on the Earth has been hypothesised it is again important to recall that any mechanism that successfully removes H_2O from the environment of Venus is likely to perform the same but unnecessary service for the Earth! Pollack and Yung (1980) point out that these different hypothetical histories should have produced different effects on the D:H ratio for Venus. There is hope, therefore, that future satellite or surface observations can resolve the problem†.

Inconsistencies in models of the atmospheric evolutionary history of Venus centre on the fact that if volatile inventories for the Earth and its 'sister' planet were indeed originally

---

† The D:H ratio observed by Pioneer Venus is currently being interpreted as indicating extensive loss of hydrogen from the planet and thus presumably from an early H_2O dominated atmosphere.

similar the mechanisms of removal and modification for one but not the other are likely to be functions of or associated with planetary differences. These are (i) planet–Sun distance, (ii) planetary rotation rate and (iii) the Earth's moon.

It is important to note here that Venus is probably the least well understood of all the terrestrial planets. At the present time there are measurements that indicate that there may be more energy being emitted by the planet than is absorbed by it from the Sun (Tomasko *et al* 1980). However further analysis of the net flux observations made by Pioneer Venus suggest that this initial value is too high (Taylor *et al* 1980, Taylor 1981).

The temperature history of the planet Mars is clearly going to be much more difficult to treat as a smoothed long-term evolution. The surface features observed by Mariner 9 and Viking spacecraft seem to be the result of fluctuating levels of atmospheric and tectonic activity (e.g. Guest *et al* 1979). The sub-evolutionary cycles in the martian 'climate' (§5.2) are of more importance and interest than the long-term trends which depend only upon known levels of outgassed CO_2. Even these levels are difficult to estimate when the anomalous $^{15}N:^{14}N$ ratio (e.g. McElroy *et al* 1977) are considered. Pollack and Black (1979) have used the spacecraft data to estimate that the total outgassed inventory for Mars represents only about 20% of the available volatile content. It should be noted here that scaling with ^{40}Ar presumes a knowledge of the comparative planetary outgassing rates since ^{40}Ar is only released during outgassing. Since, as described in §2.1, the tectonic histories of the terrestrial planets are likely to have followed a similar sequence but over and at different time periods a direct scaling may provide, at best, only a first-order estimate of available volatiles. The existence of water vapour in the martian atmosphere is a necessary consequence of the existence of surface liquid water, for which there now appears to be an ever-increasing amount of data (e.g. Guest *et al* 1979, Mutch 1979, Spitzer 1980). However, it is possible and seems probable that most, if not all, (the most questionable is the first epoch) of these epochs of enhanced climatic conditions were comparatively short lived (in a geological sense). It is for this reason that the

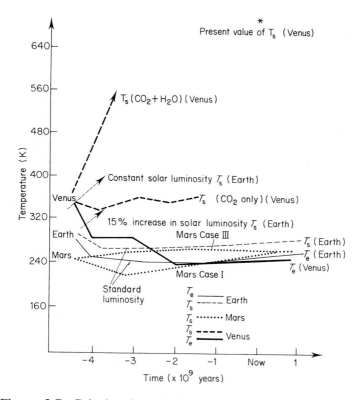

Figure 3.7. Calculated evolutionary histories for the Earth (fine lines), Venus (heavy lines) and Mars (dotted lines). The possible change in rotation rate has been omitted (cf figure 2.2) and the Earth's effective temperatures T_e, and surface temperature have been calculated for different percentage changes in incident solar radiation. The effective temperature of Venus decreases as the planetary albedo increases and the surface temperature is a function of atmospheric composition. Calculations for both Venus and Mars have been made only for the standard solar model. Surface temperatures are plotted for Mars for two cases of I early degassing and III later degassing (cf figure 5.6). For all the histories, the initial incident solar flux and the rate of buildup of the atmosphere seem to be critical.

surface temperature curve for Mars in figure 3.7 has been drawn only for a CO_2 atmosphere. The evolutionary curves of T_s illustrated in this figure, follow the type of atmosphere–surface interaction histories, under the influence of an increasing solar luminosity, as described in §3.2 and the appendix.

There are still numerous unresolved discrepancies between the models of noble gas and total volatile inventories of the three main terrestrial planets and their observable atmospheres. If Mars was originally endowed with the same proportions of volatiles as the Earth then a simple scaling analysis plus a very reduced degassing efficiency (approximately $\frac{1}{5}$ that of the Earth) which may be the result of lower internal energy sources leads Pollack and Black (1979) to predict that approximately 1000–3000 mbar of CO_2, 50–100 mbar N_2 and between 80 and 160 m of equivalent liquid H_2O will have been released during the planet's evolution, but the recent noble gas data for Venus (Donahue *et al* 1981) raise doubts about the general applicability of Pollack and Black's (1979) work and indicates that most of the outgassing took place early in the planet's history. This latter suggestion seems to be substantiated by the detailed work of McElroy *et al* (1977) on the relative abundances of the isotopic species of nitrogen. It is concluded from these studies that, currently, the outgassing must be at least 20 times lower than the average over the whole evolutionary period. This degassing history must be viewed in the light of the evidence for volcanic activity and models of the evolution of the martian lithosphere (e.g. Toksoz *et al* 1978 and figures 2.3 and 2.4).

The evaluations of total volatiles outgassed and the present atmospheric inventory must imply substantial losses of gases from Mars. Various hypotheses have been examined. These range from the computations of Yung and Pinto (1978) which indicate that photochemical processes could lead to the removal of considerable atmospheric mass as a result of the creation of hydrocarbons to the photochemical removal rate calculations of McElroy (1972) and McElroy *et al* (1977). The former are dependent upon the existence of an early atmosphere dominated by CH_4 which now seems unlikely whilst the latter lead to, at most, removal of about 3 m of H_2O and around 1–2 mbar of N_2. It seems inevitable that appeal must be made to incorporation of considerable atmospheric mass into surface and sub-surface reservoirs. For instance, Fanale and Cannon (1979) believe that the lithosphere currently contains at least 20 times the atmospheric mass of CO_2 and the Viking lander experiments (Kuhn *et al* 1979) substantiate hypotheses

116

regarding the incorporation of considerable quantities of oxygen (presumably released as a result of the photochemical break-up of H_2O) into the surface regolith. Despite these inventories and the additional surface reservoirs of the polar caps, a considerable imbalance remains between the likely outgassed volatile amount and that accounted for in the atmosphere, the lithosphere, surface reservoirs and through currently viable atmospheric escape patterns. Levine (1978) examines the volatile inventory for Mars. Photochemical dissociation and escape is likely to have accounted for only about $1-3 \times 10^3$ kg m^{-2} of H_2O and 10 kg m^{-2} of CO_2. He concludes that chemical weathering must provide a massive sink for atmospheric volatiles. This gap has contributed to the suggestions of earlier 'more clement' or at least more substantial atmospheric configurations on Mars. In view of the very low levels of solar energy at the martian orbit such hospitable conditions are likely only as part of a (possibly declining) cyclic pattern. Chapter 5 discusses such short-term planetary climatic cycles in detail.

3.4.2. *Titan*

Titan, Saturn's largest satellite, ranks second only in size to Jupiter's Ganymede amongst the solar system satellites. It has a radius of 2560 ± 26 km and the most substantial atmosphere of any satellite. The surface pressure is approximately 1600 mbar (table 3.4) i.e. the atmosphere is over $1\frac{1}{2}$ times more massive than the Earth's though surprisingly similar in chemical make-up as its predominant constituent is molecular nitrogen (table 3.6). The northern hemisphere has been shown by Voyager 1 to be darker than the southern hemisphere and there is a dark polar hood over the North Pole. (Voyager 2 observations show this feature reduced in size and relocated at approximately 65°N.) The surface temperature has now been estimated to be approximately 93 K. Figure 3.8 illustrates a probable temperature lapse rate curve for Titan. This is the model temperature variation with height used by the Voyager 1 investigating teams to interpret the infrared spectral observations (Hanel *et al* 1981). Methane probably composes approximately 6% of the atmosphere at

Table 3.6. Atmospheric composition of Titan from Voyager 1 (after Hanel *et al* 1981).

Gas	Band	Wavenumber (cm^{-1})	Approximate mole fraction or percentage
Positively identified			
Methane (CH_4)	v_4	1304	1×10^{-2}
Ethane (C_2H_6)	v_9	821	2×10^{-5}
Acetylene (C_2H_2)	v_5	729	3×10^{-6}
Ethylene (C_2H_4)	v_7	950	1×10^{-6}
Hydrogen cyanide (HCN)	v_2	712	2×10^{-7}
Tentatively identified			
Methylacetylene (C_3H_4)	v_9, v_{10}	633, 328	
Propane (C_3H_8)	v_{26}	748	
Proposed			
Nitrogen (N_2)	—	—	>90%
Argon (^{40}Ar)	—	—	up to 10%

the surface and less than 1% of the upper atmosphere (table 3.6). Complex and continuing photochemical and chemical reactions are evident. The range of organic molecules detected is wide and the detection of hydrogen cyanide (HCN) is of particular interest since this compound is the key precursor of aminoacids and bases in nucleic acids (chapters 4 and 6). The source of the HCN is probably photolysis of CH_4 in the presence of considerable N_2, although Strobel (1981) has recently suggested that CH_4 is more efficiently broken down by electron bombardment. The existence of a hydrogen torus surrounding Saturn near the orbit of Titan (similar to that of Io, figure 2.8) was first suggested in 1973 (McDonough and Brice 1973). Hydrogen atoms and molecules, probably resulting from photochemical reactions involving CH_4 can escape from the exosphere of Titan because of its comparatively low mass but they cannot escape the gravitational field of the parent planet. If the source of the neutral H observed by Voyager 1 is photolysis of CH_4 from Titan then Broadfoot *et al* (1981) calculate a necessary source strength of 2×10^{13} atoms/m^2 s^1 at a radius of 3000 km.

Figure 3.8 illustrates that the observed surface temperature

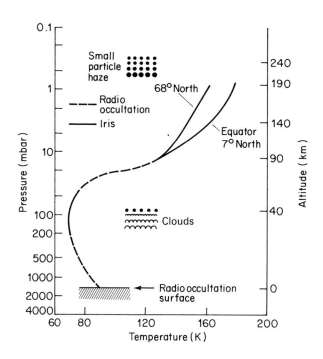

Figure 3.8. Temperature profile of Titan from Voyager 1. The full curves are from IRIS data and broken parts are from the profile of the temperature–molecular weight ratio obtained by the Voyager radio science investigation. Haze and condensation zones are indicated schematically (after Hanel *et al* 1981).

on Titan is very close to the triple point of methane (90.7 K at 117 mbar). Thus methane will play a similar role to water on the Earth. It is anticipated that on the surface, liquid and solid phases of CH_4 will exist and that clouds and aerosol formation in the troposphere will be very strongly dependent upon the vapour pressure of CH_4. The importance of CH_4 in the photochemistry of the upper atmosphere has already been discussed. If C and N have solar relative abundances then Tyler *et al* (1981) suggest that there will be approximately 100 times more CH_4 on the surface (presumably as a liquid or solid) than observed in the atmosphere. These authors follow the analogy between the Earth's water and Titan's methane to the extent of suggesting that negative feedbacks could exist between the surface temperature of Titan and the phase state of CH_4 on the surface and in the atmosphere.

The evolutionary steps which have formed an atmosphere

of molecular nitrogen on Titan have been considered by Atreya *et al* (1978). They proposed that an outgassed mixture of CH_4 and NH_3 would be altered by photolysis to N_2 and H_2. The hydrogen, methane and ammonia would contribute to a greenhouse temperature enhancement and the H_2 provides the source for the neutral H_2 torus. Radio measurements from Voyager 1 suggest a mean molecular weight of between 28 and 29. The results are not therefore compatible with a purely N_2 dominated atmosphere but suggest a small admixture of a neutral component with a higher molecular mass than 28. Argon (up to 10% in total) has been proposed (see tables 3.4 and 3.6).

Comparison between features of the atmospheric evolution of Titan and the terrestrial planets may be useful. For example, the upward movement of CH_4 is constrained by a 'cold trap' at the tropopause. The dynamics of Titan's atmosphere seem similar in some ways to those of the atmosphere of Venus. There is no observable longitudinal temperature structure and since the day length is short ($\simeq 16$ days) compared with the radiative relaxation time ($\simeq 138$ years) there is no diurnal temperature signal. The estimated wind velocities range from about 60 m s^{-1} in the troposphere to 110 m s^{-1} in the stratosphere. These very high stratospheric velocities ($10 \times$ planet's rotation rate) seem to be similar to those observed on Venus by Mariner 10 and the Pioneer spacecraft. Titan is tidally locked to Saturn and has an obliquity of approximately $27°$. Thus the seasonal change is greater than on Earth. It is possible that both the asymmetric hemispheric colouring and the hood/ring feature derive from seasonal insolation changes. It seems likely that there is a weak zonal flow pattern similar to that on Venus.

Finally there are several significant haze layers in Titan's atmosphere. These have been observed in limb images and are found to be located at heights of between 200 and 400 km above the surface. A lower limit to the haze layer optical depth can be estimated from the fact that no surface features are seen on Titan as opposed to all the icy satellites of Saturn (Smith *et al* 1981). The haze layer is quite opaque at visible wavelengths—similar to that of Mars at the height of a global dust storm.

It is fascinating to notice in figure 3.8 that the observed temperature profile on Titan resembles that of the Earth in as much as there is a significant temperature inversion in the stratosphere. This temperature increase ($\simeq 160$ K cf 80–90 K at the surface) is caused, on Titan, by the aerosol layer (possibly composed of polymerised hydrocarbons) which is optically thick to visible radiation but thin in the thermal infrared (see discussions of cloud feedback in §2.3.2). It is not unreasonable to suggest that the evolution of atmosphere and climate on Titan have been, at least partially, controlled by negative feedback effects similar to those seen on the Earth (chapter 4).

The photochemistry of the evolving titanian atmosphere may be more important than for the other terrestrial planets. The present configuration is very active photochemically with detection of gaseous phase hydrocarbons being made by both Voyager spacecrafts. Over evolutionary time periods the formation of aerosols by polymerisation of simple organics may prove to be an important sink for CH_4 since these aerosols have a theoretical atmospheric lifetime of only 1 year (Podolak and Bar-Nun 1979). There must, therefore, have been a significant mass loss to the surface over the lifetime of the solar system. Before the Voyager observations of Titan relatively little attention had been paid to the effects of photolysis of NH_3 in the atmosphere. However, Hunten (1977) had suggested that this process was likely to have generated huge amounts of free N_2. Observations seem to support this prediction but it has not yet been established whether temperatures were ever high enough ($\gtrsim 150$ K) to allow NH_3 into the atmosphere before conversion to N_2. Alternatively all the atmospheric N_2 may have originated directly by degassing. This suggestion is supported by the absence of CO and Ne in the titanian atmosphere. Further analysis of the evolutionary history of Titan would clearly have application to the rest of the solar system.

The probable evolution of Titan's atmosphere is very difficult to establish. Its earliest stages must have been strongly dependent upon the original chemical composition of Titan itself but the surface temperature will have been affected by the early high radiation flux from Saturn (§3.1) as well as the reduced solar flux. The outgassed volatiles may have included H_2O, N_2 (or possibly NH_3) and CH_4 all of which would have condensed rapidly on to the surface. The subsequent physical and chemical evolution would have been critically dependent upon the rate of outgassing and the chemical and photochemical reactions occurring. By analogy with the other terrestrial planets the questions to be posed include the rate of migration of solid state volatiles towards the polar regions; the rate and nature of loss to the surface and to space and, for instance, the time period before which the atmosphere was substantial enough to sustain the aerosol and clouds seen today (figure 3.8).

Golitsyn (1979) gives an interesting review of the characteristics of planets and atmospheres which determine the nature of their dynamics. Table 3.7 lists some of these atmospheric parameters. The final parameter is the ratio of the time required to attain local thermodynamic equilibrium to the radiative cooling time, τ_r. If this parameter is much less than unity the atmosphere is far from radiative equilibrium, being dominated by large scale dynamic motions whereas if the value is greater than 1 then the atmospheric state is much more closely associated with radiative fluxes. The interesting feature here is that Titan falls into the same category (dynamic) as the planets (excluding Pluto) whilst the other satellites seem to show a regime dominated by radiative rather than dynamic equilibration. This seems to justify further the inclusion of Titan among the class of planets rather than satellites.

3.4.3. Triton

The inclusion of the satellite Triton in this subsection is based upon two confirmed observations of spectral features consistent with an atmosphere. The size of Triton (table 3.4) and its general similarity to the other major satellites at least suggests that it could be a candidate for a role as a planet with an atmosphere. However until 1979 ground based observations had resulted in either negative results (Martin 1975, Cruikshank and Silvaggio 1979) or uncertain features (Fink et al 1977). Cruikshank and Silvaggio (1979) report a

122

Table 3.7. Atmospheric parameters for selected planets and satellites (after Golitsyn 1979). Note that Titan resembles a 'planet' more closely than Pluto and the other tabulated satellites.

Object	p_s (Pa)	μ	C_p (10^{-2} J kg^{-1} K^{-1})	γ_a (K km^{-1})	T_e (K)	H (km)	c (m s^{-1})	τ (s)	Similarity parameter Π_M
Earth	10^5	29	1	9.8	255	7.3	270	10^7	1.2×10^{-3}
Jupiter	7×10^4	2.2	13	2.1	130	20	700	3×10^8	3×10^{-4}
Saturn	10^5	2.2	13	0.72	70–97	33	590	2×10^9	5×10^{-5}
Uranus	10^5	2.2	13	0.67	57	25	460	10^{10}	4×10^{-6}
Neptune	10^5	2.2	13	0.9	45	15	380	3×10^{10}	5×10^{-6}
Pluto	~10	16	1.8	0.8	50–63	50	180	10^6	10^{-2}
Io	10^{-5}	30	1	1.8	96	15	190	0.1	10^5
Ganymede	0.1	30	1	1.3	109	20	200	1200	10
Titan	10^3–10^5	16	1.8	0.7	84	30	250	10^{10}	3×10^{-6}

conclusive measurement of a 2.3 μm absorption feature which they attribute to gaseous CH_4. Their spectral analysis leads them to suggest that the partial pressure of methane at the surface of Triton is $1.0 \pm 0.5 \times 10^{-4}$ bar ($\approx 7 + 3$ m atm). This value is consistent with the probable surface temperature of about 55 K (Cruikshank and Silvaggio 1979). More recently Johnson *et al* (1981) have reported observations which suggest a smaller surface pressure of CH_4 of approximately 1 m atm. Cruikshank and Silvaggio (1979) failed to find evidence for surface frost deposits of CH_4. At these temperatures atmospheric CH_4 would be expected to be in equilibrium with solid surface deposits and observations of relatively high surface albedos ranging between 0.2 and 0.4 also indicate surface frost. They therefore postulate that there are surface deposits of solid CH_4 on Triton but that these are currently hidden from the surface observers, possibly because the dark (night) side of Triton could be acting as a cold trap. The slow but variable geometry of the configuration of Neptune, Pluto and Triton could cause long-term migration of considerable quantities of CH_4 over Triton's surface in a manner somewhat analogous to the movement of CO_2 on Mars (see chapter 5). There is no evidence for any other gases in Trition's atmosphere (Trafton 1981).

Discussion of the solar system bodies with both surfaces and significant atmospheres indicates that a number of features seem to be necessary for, or at least intrinsic to, any persistently stable configuration. The history of the Earth's atmosphere offers the opportunity to explore some of these internal feedback effects between planet and atmosphere more fully.

4. The Earth's Atmosphere and the Origin and Evolution of Life

The chemical composition and physical state of the Earth's atmosphere is the result of the complex interplay of processes acting throughout geological history. Two very important properties of the Earth's atmosphere–climate system identify it as different from all the other planets. These properties are the existence of water in its liquid state on the surface and, at the same time, in its vapour state in the atmosphere and the presence of life. These two properties have led to considerable modifications in the atmospheric and climatic regimes. The presence of liquid water has permitted the removal of CO_2 from the atmosphere and its deposition, in the form of carbonaceous sedimentary material, into the crust. (table 2.5 and figure 3.4). The magnitude of this effect is clear when the total atmospheric constituents of the Earth and Venus are compared (table 1.5). The atmospheric effects which control surface conditions making the Earth hospitable for the existence of life may be listed as follows:

1. greenhouse enhancement of temperature;
2. absorption of high-energy ultraviolet radiation;
3. circulation of energy resulting in an equable climatic regime at most surface locations;
4. 'moist' systems (i.e. when an atmospheric constituent changes phase readily releasing/removing latent heat) have a large capacity to dissipate energy imbalances.

These crucial effects are readily seen to be the result of the presence of the hydrosphere and biological activity though

125

the presence of atmospheric oxygen, and hence O_3 in the stratosphere, may predate the origin of life. The earliest stages of atmospheric evolution following the formation of the Earth may therefore be critical (cf §3.4). Subsequent atmospheric evolution must be viewed as a component part of planetary evolution which, for the Earth, included the origin and evolution of the biosphere.

It is not certain that life originated on the Earth. There have been a number of suggestions in the literature relating to the concept of panspermia (e.g. Hoyle and Wickramasinghe 1979). Certainly there must have been a considerable input of simple organics by accretional and meteoritic bombardment of the early Earth (e.g. Henderson-Sellers *et al* 1980, Schwartz 1981).

Cometary impact has been suggested as the source of both life itself and, perhaps more interestingly, of volatiles and organic compounds (table 3.5). Irvine *et al* (1981) list the atomic and molecular species observed in comets (see table 4.1). These authors make the important point that observation of species in the coma does not necessarily imply the existence of the species or its parent molecule in the cometary nucleus. A careful consideration of the gas phase chemistry is required before any such hypothesis can be made. Neither HCN nor H_2O have been observed directly although there is strong circumstantial evidence which supports the view that at least some comets are predominantly H_2O. Amino acids (Kvenvolden *et al* 1970, Cronin *et al* 1979) and the bases of nucleic acids (Stoks and Schwartz 1981) have been found in carbonaceous chondrites (see table 4.1 and discussion of table 3.5) and it seems plausible that in comets such molecules may also exist. Unfortunately the

Table 4.1. Atomic and molecular species observed in comets (after Irvine *et al* 1981).

Organic:	C, C_2, C_3, CH, CN, CO, CS, (HCN), (CH_3CN)
Inorganic:	H, NH, NH_2, O, OH, Si, (H_2O)
Metals:	Na, Ca, Cr, Co, Mn, Fe, Ni, Cu, V
Ions:	CO^+, CO_2^+, CH^+, CN^+, N_2^+, OH^+, H_2O^+
Dust:	Silicates

effect of a cometary collision with the Earth is likely to be considerably more traumatic than that resulting from meteoritic impact, both for the impacting material and for the Earth itself. The possible perturbations caused by cometary collision on a planetary environment are discussed in chapter 6. Lazcano-Araujo and Oro (1981) have investigated the chemical reactions which may occur in the highly disrupted region of the atmosphere immediately following a cometary impact. It appears that the conditions are similar to those of a Fisher–Tropsch reaction and that a number of gas phase products of biological significance can result. Reservations regarding this method of deposition of organics are that it depends upon the appropriate timing of an impact and the subsequent transport of the organics to the surface. This latter problem is common to all high-atmosphere gas phase reactions invoked for the formation of pre-biological molecules.

A number of impact events involving large meteorites have been documented. There are also suggestions that larger, asteroid-sized bodies or comets have impacted the Earth. Alvarez *et al* (1980) have mapped the locations of iridium-enriched deposits around the world. In at least two areas of the globe (Denmark and Umbria) the Ir is found to occur almost precisely at the Cretaceous–Tertiary boundary. Alvarez *et al* (1980) suggest that this iridium is of extra-terrestrial origin and probably indicates the collision of a large body with the Earth around 65×10^6 years BP (see the discussion of shorter-period perturbations in chapter 5) and may be associated with the extinctions known to occur at this boundary.

Despite the fact that comets are believed to be rich in volatiles, it seems likely that a 'dead' comet (i.e. a comet from which most of the volatiles have been driven and which therefore does not exhibit tails as it approaches the sun) would lose many of these volatiles within a few orbital periods (the order of hundreds of years) whereas the time period before collision could be as long as 10^8 years (Wetherill 1975, 1980). However even 'dead' comets are likely to be rich in organic compounds although the volatile content will have been severely reduced. These organics would almost

127

certainly survive and probably metamorphose under UV irradiation at and following impact.

The problem of transfer of biochemically interesting material from space to the Earth's surface has recently been considered by Irvine *et al* (1981). These authors calculate that the Earth accretes approximately 10^4 tonnes of micrometeoroidal material each year. Since the source of much of this very small meteoritic material is believed to be cometary debris and because destructive thermal effects decrease with the decrease of particle size the resultant flux of uncontaminated cometary material could be substantial. Such a process has also been suggested by Hoyle and Wickramasinghe (1979) in their more extreme hypothesis of panspermia (chapter 6).

Table 4.2 lists some major characteristics of the planet Earth. The suggestion that climatic stability is a property peculiar to the Earth will be examined here and in chapters 5 and 6. At each stage in the evolution of the Earth, from the earliest bombardment of the planetary surface by planetesimals (see §§2.1.1, 2.1.2 and figure 2.1), through short period extreme events such as cometary impact (chapter 5) and within the complex atmosphere–surface cycles, the importance of interactions in and between the biosphere and climate will be apparent. There is a second facet of the atmosphere–biosphere system which will be developed in this chapter: the effects of the biosphere upon the atmosphere. These two extremes are clearly not independent of one

Table 4.2. Some fundamental characteristics of the planet Earth (modified from McCrea 1981). 'A strange and beautiful anomaly in our solar system' (Lovelock 1979).

Orbit:	obliquity considerable, eccentricity significant
Interior:	still active
Magnetism:	relatively strong, variable
Crust:	non-rigid and non-static
Hydrosphere:	unique
Atmosphere:	numerous unique features
Biosphere:	unique
Satellite:	largest in relative mass
? Climate:	stable (unique?)

another nor of the other external and internal atmospheric forcing effects which are described in other chapters. The interaction of the biosphere with the other components of the Earth's environment may be of fundamental significance for the stability of these systems (Lovelock 1975, 1979).

Any discussion of the origin and early evolution of life on Earth must consider the composition of the atmosphere. This is because the atmosphere plays a major, and often dominant, role in shaping the chemical and physical environments found on the Earth. Even though detailed consideration of the nature and timing of evolutionary processes will probably be focused upon a specific *surface* location, it is the properties of the atmosphere which define the surface conditions. The type of surface is, very often, a result of atmospheric processes. Just over 70% of the Earth's surface is covered by water†; also desert and ice surfaces are a result both of specific location and processes of global scale circulation (see figure 2.4).

Photosynthesis and respiration depend directly upon the chemical composition of the atmosphere‡.

The features of the atmosphere to be considered are
    (i)   the total atmospheric mass,
    (ii) the chemical composition, and
    (iii) the temperature lapse rate within the troposphere.
Together these parameters will permit definition of the likely average and possible extreme environmental constraints around the globe. The three topics are clearly not independent. The example from the present-day literature of the CO_2 budget (see chapter 2 and figure 2.13) will serve to illustrate some of the complexities which relate the mass, composition and temperature regimes.

4.1. The Pre-biotic Environment

Geological data are extremely sparse for most of the early (say older than 3.2×10^9 years BP) Precambrian epoch. An excellent review of the data and theories pertaining to the

---

† As A C Clarke has observed: 'How inappropriate to call this planet Earth when clearly it is Ocean'.
‡ Species adaptation to particular ecological niches is obvious in the variation of partial pressures of both O_2 and CO_2 to which biota respond.

evolution of the Earth's crust is given by Goodwin (1981). It is extremely difficult to establish the sequence and time span of the events which contributed to the formation of the lithosphere and the hydrosphere (figure 2.1). Although the primitive atmosphere is unlikely to have been significantly modified by life, its interaction with the juvenile lithosphere and hydrosphere may have been crucial. The development of the Earth's crust must have been the result of the complex interplay between a vigorous mantle circulation, accretional bombardment and energy and chemical transformations resulting from interactions caused by input of hot magma into the young surface environment. The schematic diagram of crustal evolution in figure 2.3 suggests a period of early 'scum' tectonics in which very thin crustal plates more like the oceanic crust than continental plates were being rapidly created and destroyed. In Moorbath and Windley (1981) it is suggested that in the case of the Earth this rapid destruction of crustal material was even further increased by the 'lubricating' effect of liquid water which assisted subduction of these thin plates (cf Goodwin (1981) in which the establishment of more than 10% crust is dated at between 3.0–2.5 × 10^9 years). The temperature profile through the uppermost layers of the lithosphere appears to be very important since it is this that determines not only the chemical and physical nature of the rocks formed but also the flux of energy to the juvenile atmosphere and hydrosphere (figure 2.1). Clearly the energy input could never have become large enough to cause the vaporisation of the whole hydrosphere. This upper bound to the energy flow does not seem to be incompatible with a period of intense tectonic activity particularly if the upwelling magma were localised. A further constraint from the geological data is that the earliest known rocks include metamorphosed *sediments*. Thus it seems very likely that land areas existed before the time at which this sedimentation began. Since the Isua rocks are now believed to be 3.83 × 10^9 years old (Moorbath *et al* 1973, Schidlowski 1980a), the terrestrial environment must have included land surfaces before this time. These tenuous lines of reasoning seem to lead to a picture of the earliest stages of environmental evolution on the Earth which includes both a rapidly cycling thin oceanic crust and a larger number of volcanic islands.

These volcanic islands would provide the land surface for erosion of sedimentary material throughout the period before continental crust was evolved. Such islands probably existed in groups centred on a site of large-scale mantle upwelling. Moorbath (1982) believes that although their existence may have been short lived the total surface covered by these islands could have been considerable (see also Moorbath and Windley 1981, Goodwin 1981).

A stable planetary environment and, probably, a crust seem to be important for the origin of life. Chemical interactions may have occurred in surface situations, possibly using clay catalysts, and/or at the site of submarine hydrothermal vents (see e.g. Corliss *et al* 1981). Experimental results suggest that organic–mineral interaction is the most satisfactory and possibly the only route for synthesis of polypeptides. The work of Lahav *et al* (1978) and Schwartz (1981) and others has established the importance of a surface environment including both land and water areas and changing (say in a daily cycle), but not highly perturbed, conditions.

The bombardment of the surface is likely to have been less destructive on the Earth once an atmosphere was established (see Benlow and Meadows 1977). However, if deposition of organics already formed in space is likely then the timing of bombardment and its possible effects upon the environment must be considered (figure 2.1).

Crustal growth continued throughout much of the Archaean (3.8–3.0 to 2.5×10^9 years BP) and at the end of this period between 40% and 75% of the present-day continental crust had been created (Windley 1980, Basu *et al* 1981). Windley (1980) suggests that this very high rate of growth was a result of heat supplied by the decay of radiogenic isotopes (in the modern environment over 70% of the Earth's internal energy is utilised in plate formation, see figures 2.1 and 2.3). However the rate of crustal growth is not yet established. McCulloch *et al* (1981) suggest a large increase in crustal formation at 2.6×10^9 years BP. By 2.0×10^9 years BP it seems likely that between 80% and 90% of the crust had been formed.

In any description of the evolution of life and the atmosphere adequate definition of the timescales over which mechanisms operate is crucial. In this discussion of the Earth's atmosphere the initial analysis will be restricted to the time period over which the evolution of *life* and the evolution of the *environment* coincided. This is the most exciting period in terms of atmospheric evolution because it is during this time that atmospheric and surface properties define the regimes within which evolution must progress, whilst at the same time the enormous strides of evolutionary processes may impinge upon, or even dominate, global conditions.

The time period of maximum interaction (see figure 4.1) lies between the earliest period of planetary formation termed the astrophysical environment and the period around the mid-Precambrian when, after the appearance of significant free oxygen in the Earth's atmosphere, the regime may be termed the modern environment. Positioning on the time axis of even these end points is difficult. The start of the modern environment probably lies between 2.5 and 1.6×10^9 years BP (Levine *et al* 1979a) although the timing and even the definition of an oxidising environment is difficult (e.g. Kasting and Walker 1981). At the other extreme the very earliest history of the terrestrial environment must

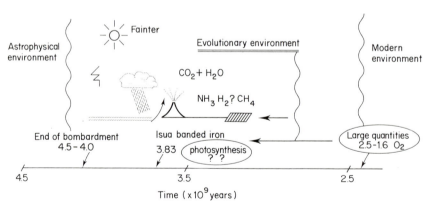

Figure 4.1. Schematic diagram indicating the time span covered by the *evolutionary environment* which extends from the cessation of planetary-scale astrophysical upheavals until the geological era when the modern ocean/continental ratio is established and oxygen is a significant constituent of the atmosphere (after Henderson-Sellers 1981, 1982).

132

be dominated by processes associated with planetary formation. Bombardment by debris ranging in size from meteoritic to interstellar dust (§2.1.1), sweeping by the solar wind, subsurface heating through the release of radioactive energy and the physical and tectonic upheavals of core and mantle formation and gravitational attractions say between the Earth and proto-Moon and Venus suggests an environment likely to be inhospitable for life. The cessation of these major disruptions is again very difficult to locate in time. Certain models of planetary formation can build the Earth and its atmosphere within 10^5 to 10^6 years whilst other evidence (such as that from the history of lunar surface bombardment and disruption) may indicate a much longer process (Williams 1975 and chapter 1). Hence the start of the time period of interest lies between 4.5×10^9 and approximately 4.0×10^9 years and its termination between approximately 3.9×10^9 and 2.5×10^9 years BP.

It is within this time period, which defines the evolutionary environment (Henderson-Sellers 1981) and has a suggested length of between 400 and 600 million years (see figure 4.1), that the interactions between the evolutionary processes of life and the composition of the atmosphere take place (see also figure 2.1(*a*) and Goodwin 1981).

4.2. The Evolutionary Environment

Since by definition there are very few data for this period the discussion must of necessity be somewhat tentative. It is therefore extremely important to list as many relevant data as possible. These data will be both observations and results from theoretical models. The evolutionary environment has the following characteristics (Henderson-Sellers 1981).

4.2.1. A CO_2 and H_2O Vapour Atmosphere Containing Other Neutral and Trace Gases

Current thinking about the early atmosphere of the Earth (and indeed all the terrestrial planets) supports the opinion (e.g. Abelson 1966) that the primitive atmosphere was a nearly neutral or mildly reducing mixture of the volatiles CO_2, H_2O, N_2 and CO (Walker 1977, Owen *et al* 1979,

133

Henderson-Sellers *et al* 1980). The propositions of Owen (1978) relating to terrestrial planetary atmospheres are listed in §2.1.1. Geophysical evidence for the oxidation state of the early atmosphere can be interpreted in a number of ways and the sparseness of data from the mantle makes unequivocal statements impossible. Arculus and Delano (1980) have reported oxygen fugacities in mantle-derived spinels which imply a highly reduced upper mantle. It can be convincingly argued (e.g. Owen 1978, De Paulo 1981) that the atmosphere and hydrosphere were derived from a volatile-rich veneer accreted after core–mantle separation. Alternatively Davies (1981) suggests that the mantle composition is heterogeneous and therefore individual observations are difficult to interpret in terms of the redox state of degassed volatiles. The argument of Walker (1977), which followed Hanks and Anderson (1969), suggests that the early atmosphere was unlikely to be more than very weakly reducing because it was derived from a mantle that lacked free metallic iron. This can be seen to fit new geological data if the core separated steadily as the Earth accreted (Cogley 1982). Large partial pressures of NH_3 and CH_4 cannot persist in an environment dominated by a global hydrosphere (see §4.2.4). Kuhn and Atreya (1979) have demonstrated that NH_3 is lost rapidly while Levine and Augustsson (1981) have performed simulations which permit the estimation of the lifetime of CH_4 in the atmosphere. They calculate that this value is about 50 years (see §4.2.7). It is by no means certain precisely how or at what rate the atmosphere evolved. The rate of atmospheric build-up is discussed in §2.1 and below. However, the time span of the evolutionary environment is such that chemical and geophysical processes will warrant detailed consideration. The removal of gases (e.g. CO_2) from the atmosphere not only affects the total atmospheric mass but also decreases the greenhouse effect. Chemical reactions (especially photochemical ones) may considerably perturb levels of trace gases; small amounts of H_2, NH_3 and CH_4 may have been present.

4.2.2. *An Atmospheric Mass of Between 500 and 1000 mbar*

The effects of an atmosphere on surface temperature depend on the amount of atmosphere present, as well as on its chemical composition; but the former factor depends on the (highly uncertain) degassing rate. Noble gas data from the Viking mission to Mars reveal a very similar pattern of relative abundances as found on the Earth. These ratios are also found in the 'planetary component' of meteoritic gases reinforcing the view that fractionation occurred prior to planetary formation. The classic work of Rubey (1951) is supported by a comparison between the total volatile inventory of the Earth and the best estimate of the Venus inventory from Venera 9 and 10 (see table 1.5, after Owen 1978), although Pioneer Venus data (e.g. Donahue *et al* 1981) indicate that the cytherean ratios differ substantially (see chapter 3 and figure 3.5). Degassing/build-up rates for any of the terrestrial planets are uncertain. For the Earth the modern degassing rate is non-zero for at least two of the noble gases, xenon and helium, but the rate of addition of volatiles is likely to be negligibly small. The atmosphere has probably been separated from the upper mantle for approximately $4.4–4.5 \times 10^9$ years (Arculus and Delano 1980). There is, at least a lower bound which can be placed upon the atmospheric pressure once a hydrosphere exists, i.e. the partial pressure of water vapour is approximatably 6 mbar. The geological data implying a deposition of sedimentary material in either an air or water environment also require minimum levels of surface pressures although these are much less easy to evaluate. More convincing is the argument that in the highly active tectonic and bombardment environment the build up of pressure would be rapid. If the models of Fanale (1971) can be substantiated then catastrophic degassing could give surface pressures considerably higher than present-day values. The ambient pressure would then depend upon the rate of removal of atmospheric species.

In view of the considerable uncertainties Henderson-Sellers *et al* (1980) have looked at two extreme cases: (1) when the degassing rate is effectively instantaneous; and (2) when the degassing rate is slow (i.e. taking 10^9 years or more to produce an atmosphere nearly equal in amount to that present now). Slower degassing than that taken in (2) would lead to some conflict with current geological evidence, especially with the argon isotope data. Figure 4.2 illustrates the

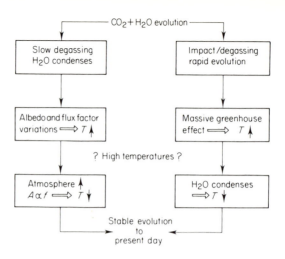

Figure 4.2. Two extreme evolutionary histories for the Earth's atmosphere—in both cases condensation of surface liquid water occurs and the surface temperatures stabilise (after Henderson-Sellers and Meadows 1978).

two extreme evolutionary histories described in chapter 2. Both histories (it is important to note that the actual evolution probably followed a path between these two extremes) suggest a possible enhancement of temperatures (figures 2.2 and 3.7). The surface temperature of the Earth, according to both models, has always been between the freezing and boiling points of water.

Sedimentary carbonate rocks are present throughout the geological column. This limits the atmospheric composition a little since it implies the existence of some atmospheric CO_2 throughout the period of the established geological record (i.e. at least 3.8×10^9 years). Holland (1978) has suggested that the early Precambrian rocks seem to be enriched in dolomite $(CaMg(CO_3)_2)$ compared with the more modern predominance of calcite $(CaCO_3)$. This, he believes, is related to higher heat fluxes through the mantle in these periods. Holland (in Pollack and Yung 1980) also recognises that the geological data provide an upper limit to the partial pressure of atmospheric CO_2 on the Earth since the absence of carbonates containing large amounts of Na and K implies partial pressures of $CO_2 \lesssim 100$ mbar.

4.2.3. Lower Incident Solar Flux

Current models of stellar evolution agree that the early Sun emitted appreciably less flux than the present-day Sun. For the Earth, the difference seemed sufficiently large to have produced temperatures below the freezing point of water thus conflicting with known geological data and stimulating a number of early models of the primitive atmosphere (Rasool and De Bergh 1970, Sagan and Mullen 1972). The emitted flux has probably increased by between 35% and 45% over the lifetime of the solar system (Haselgrove and Hoyle 1959, Newman and Rood 1977), see §1.2.

During the probable lifetime of the evolutionary environment an upper limit on the change in solar flux is approximated by a linear increase of the order of 4–5%. A secondary but important effect for the system evolution will be the changing nature of absorbed and emitted radiation within the atmosphere. In particular, the absorption of UV radiation, by trace gases within the troposphere (see §4.2.7) and eventually by the evolving ozone layer in the stratosphere, may affect surface and atmospheric temperature profiles.

4.2.4. Extensive Surface Liquid Water

The amount and, possibly still more important, the phase state of water in the evolutionary environment appear to be fundamental.

The geological record provides some data here. The existence of significant expanses of liquid water on the terrestrial surface can be established throughout the geological column, thus setting a long-term limit on changes in surface temperature. The presence of sedimentary rocks indicates the existence of stable regions of liquid water, since these are required for the initial sedimentation. Metamorphosed sediments are amongst the oldest rocks yet discovered (Moorbath *et al* 1973). The presence of liquid water can be established back through more than 80% of the Earth's history and may be found to extend even closer to the beginning, if older rocks can be identified. Hence, a contradiction is apparent. On the one hand, the lower solar flux suggests that water on the early Earth should have been frozen. On the other, the

137

geological evidence indicates that water was present in liquid form. This 'faint sun-enhanced surface temperature' paradox has been examined in the literature. It appears that the initial conditions that permitted condensation of degassed or vaporised H_2O may have literally 'set-the-scene' for the whole of the rest of the evolution of the Earth. If, as seems to be indicated by the geological record, it becomes increasingly clear that the depositional environment has remained singularly free of large scale changes (Holland 1972), the case for extensive (say 60–70% of the Earth's surface) palaeooceans will be strengthened. This is not only because deposition of sedimentary material requires primitive oceans similar to the present day but also, and possibly more importantly, because weathering, mass transport and particularly chemical reactions on a global scale almost certainly demand global scale[†] palaeooceans.

The stratification of the primitive ocean has also been considered by a number of authors (e.g. Henderson-Sellers and Henderson-Sellers 1978). This is of interest mainly through its effects upon the atmospheric levels of gases controlled by solution in water (CO_2, O_2, etc) but may also be important for considerations of meridional heat transport and possibly even for concentration of simple organics (Schwartz 1981). A global ocean as suggested by Hargrave (1976), however, seems to be difficult to reconcile with both the geological and pre-biological requirements for terrestrial locales (Windley 1976).

The extensive nature of surface liquid water on the primitive Earth has a number of important consequences for conditions within the evolutionary environment. (There is no difficulty in supplying adequate water at any stage during the time period of interest, see table 1.5.). The most important consequence is that a global hydrological cycle not too dissimilar from that of the present day would inevitably operate.

Extensive and/or highly asymmetric global oceanic configurations can be of considerable importance in surface temperature estimation. Ocean currents at some latitudes

† It is important to distinguish between a 'global scale ocean' which is taken here to be a set of interconnected large ocean basins of similar dimension and total area as those on the present-day Earth and a 'global ocean' which covers the whole planetary surface.

currently achieve as great a poleward energy transfer as the atmosphere (Oort and Vonder Haar 1976).

Satellite observations indicate that the present-day amplitude of the seasonal cycle of surface temperature in the northern hemisphere is about 14 K whilst in the southern hemisphere the larger ratio of water to land and hence the larger thermal inertia has reduced the cycle to about 6 K.

4.2.5. *Average Global Temperatures Between 275 K and 315 K*

There are now important observational limitations on the ambient surface temperature of the early Earth. Geological evidence indicates a global hydrosphere at least 3.8×10^9 years ago. Furthermore, the inference that is drawn from study of the metamorphosed sediments in the Isua rocks is that environmental conditions and particularly the nature of weathering and deposition 'were remarkably like those prevailing during the rest of the Archaean' (Schwartz 1981). Recent calculations using a $1D$ climate model (Henderson-Sellers and Cogley 1982) have shown that surface temperatures remain above 270 K under all likely conditions. The rest of the Precambrian geological record indicates a continuation of this apparent stability in the depositional environment (Holland 1972), although the fragmentary evidence (Frakes 1979) indicates that climatic variability was at least as pronounced as during more recent geological eras (figure 5.1). The long-term temperature record seems to agree well with the calculated temperature curves presented for the Earth in figures 2.2 and 3.7.

It seems reasonable that the oceanic extent was at least as great as the present-day extent and that the atmospheric mass was at least 300 mbar and probably considerably larger. One might therefore assert (Barron *et al* 1981) that catastrophic climatological change in the evolutionary environment will be opposed by, for instance, poleward heat flux within both the atmosphere and oceans (see also chapter 5). The poleward heat flux is usually considered to act as a negative feedback. For example, if the polar regions cool or the tropics get warmer, an increased poleward flux of heat is believed to arise, thereby driving the system towards its previous state (see §2.1.3). It is not clear how strong this climatic restoring force is, although the increased poleward heat flux in winter relative to summer provides evidence that it does exist. It is not impossible, however, to imagine situations in which circulations could act to amplify temperature changes: standing planetary-scale waves (§2.1) in the atmospheric circulation may provide a method of reinforcing and propagating surface disturbances (especially sea surface temperature anomalies).

These data indicate that surface temperatures on the Earth have been subject to short-term fluctuations but long-term 'restoring' forces seem to have been invoked which permitted continuing global stability. The nature of this mechanism is considered in chapter 5.

4.2.6. A Convectively Active Troposphere

The geological data and the model results reviewed above indicate mean temperatures around 290–300 K (in contrast with the results of Knauth and Epstein (1976)), plentiful surface water and a substantial atmosphere. The discussions of radiative equilibrium in chapter 2 lead to the expectation that convection would be important within the troposphere of the Earth. All likely atmospheric configurations would result in unstable lapse rates and thus convective mixing throughout the troposphere (see §2.2.1). Evaluation of the likely tropospheric lapse rate in the evolutionary environment is important because it permits discussion of loss through the tropopause to the stratosphere and cloud condensation. The realisation of the importance of the tropopause temperature is a result of the work of Hunten (1973) which demonstrated that the rate of hydrogen escape from the present-day atmosphere is directly controlled by the water vapour mixing ratio in the stratosphere which in turn is a function of the tropopause temperature (§2.2). Walker (1978b) has computed oxygen mixing ratios in the pre-biological atmosphere as a function of tropopause temperature (see §§2.2.1, 2.3.2). An increased value of T_{top} results in a larger net oxygen source and hence a greater mixing ratio of O_2 at the ground. Thus the evaluation of T_{top} will permit at least a preliminary attempt to integrate photochemical progression with physical and biological evolutionary processes. A first-order estimate of the cold trap temperature will be of importance for estimating mixing ratios of oxygen in the evolutionary environment (see e.g. Kasting and Walker 1981).

The discussion of convective activity and cloud formation in §2.3 has ignored the possible feedback effects of the cloud formation itself upon surface and atmospheric temperatures except as a result of albedo perturbation. Schneider (1972) demonstrated that a slight change in either percentage cloud cover or effective cloud deck height would result in a perturbation of the average surface temperature (see figure 4.3). Thus the cloud albedo feedback discussion should be complemented by consideration of the effect of likely changes in the type and/or height of cloud layers. For instance if, as has

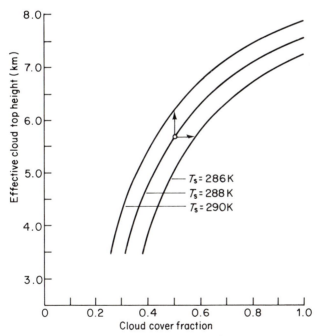

Figure 4.3. Cross plot of calculated surface temperatures as a bi-dimensional function of percentage cloud cover and effective cloud deck height (after Schneider 1972).

been suggested above, an increased value of T_s results in greater convective cloud formation (and thus a decreased total percentage cloud cover) the infrared flux emitted to space will also be perturbed. A possible feedback cycle resulting from enhanced surface temperatures as a result of an increased greenhouse effect may have to include enhanced convection; resulting percentage cloud cover; type and depth of cloud layer; convectively readjusted values of T_s and T_{top} and possibly (if temperature enhancement and/or convection processes are great enough) a deeper mixed layer and a new value for the tropospheric lapse rate.

4.2.7. Trace Gases both Abundant and Variable

The evolutionary environmental atmosphere is likely to be dominated by CO_2 and nitrogen (see §4.2.1) with a mixing ratio of water vapour determined by average surface temperatures. This assertion does not imply that trace gases are unimportant. It may be anticipated that minor constituents were both common and variable in relative and absolute abundances. Attention must be paid to these minor constituents for a number of reasons. The complex interrelationship between surface temperatures and all the other environmental constraints will be further perturbed by the presence of any atmospheric gas. The effect of additional neutral gases can be neglected in view of the results listed in

Table 4.3. Greenhouse increments (ΔT, K) caused to surface temperatures by the addition of CO or CH_4 to model atmospheres. Calculations relate to the case of radiative equilibrium (i.e. without a convective adjustment) (after Henderson-Sellers and Meadows 1979b).

Model atmosphere	Added gas (mbar)	ΔT (K)
20 mbar CO_2	0.001 CO	0.0
20 mbar CO_2	0.01 CO	0.1
100 mbar CO_2	0.01 CO	0.1
100 mbar CO_2	1.00 CO	0.6
Present day Earth's	0.001 CH_4	0.1
atmosphere	0.01 CH_4	0.7
	0.10 CH_4	2.6
	1.00 CH_4	5.8

tables 2.6 and 4.3 relating to the broadening effects of nitrogen in the atmosphere. (A comparison of these tables and table 4.4 indicates the importance of including a convective adjustment in the calculation of temperature given in these tables.) Any trace constituent possessing infrared absorption bands may be of importance. The magnitude of the further greenhouse increment is a function of the position and strength of the additional spectral features (§2.2).

A useful model atmosphere is stage 2 from Holland (1962). Here the gases are N_2, H_2O, CO_2, SO_2 with only very small amounts of H_2. Originally Holland (1962) suggested ratios of

$$H_2O/H_2 = 105$$

and

$$CO_2/CO = 37.$$

In a more recent survey (Holland 1978) these ratios have been approximately halved, giving $H_2O/H_2 = 60$ and $CO_2/CO = 17$.

Physical formation mechanisms (vaporisation and/or degassing) and geochemical equilibration at ambient surface temperatures in the presence of water lead to a neutral or mildly reducing atmosphere with almost negligible amounts of trace constituents e.g. NH_3 (i.e. less than 10^{-4} atm). However, chemical and photochemical reactions may be of local and even global importance for these trace components. Furthermore the extensive abiological and biological reactions associated with the origin and subsequent evolution of life may have considerably perturbed environmental levels of certain of these components (Toupance *et al* 1978). Interest here is in the climatological effects that these trace components may have but it is important to note that constituents such as H_2S and CH_4 may perform other significant roles—specifically the absorption of high-energy ultraviolet radiation.

Ultraviolet radiation seems to be favoured by a number of researchers in the field of pre-biotic synthesis. However, primary photodissociation reactions depend upon specific molecular absorption. This illustrates an important problem which can only be tackled by complementary work in the fields of photochemistry and tropospheric physics e.g. the

143

absorption by methane is only significant at wavelengths shorter than 145×10^{-9} m and therefore water vapour in the troposphere would effectively screen CH_4 from ultraviolet photolysis. The likely penetration of H_2O into the upper atmosphere may be of critical importance for photochemical reactions although CH_4 is likely to play a role in photochemical reactions even if it is not the primary photodissociant (e.g. Levine and Augustsson 1981).

Carbon Species. Methane is produced in the biosphere of the Earth as a result of anaerobic fermentation in soils and oceans and as a by-product of animal digestive processes. It may also be degassed at the site of mid-oceanic ridges (Corliss *et al* 1981). It persists as a trace gas in the Earth's atmosphere (Enhalt 1974). The main methane absorption occurs in two bands at 3.3 and 7.7 μm. Computations have been carried out at methane partial pressures ranging from 0.001 to 1 mbar with various values of the total atmospheric pressure. The increases in surface temperature are more significant here than for carbon monoxide (see below). This is illustrated by indicating what happens when the methane mixing ratio is varied in an atmosphere equivalent to that of the present Earth (i.e. 1000 mbar surface pressure) (table 4.3).

It is possible to argue that carbon monoxide was more abundant in the early terrestrial atmosphere than it is at present (Holland 1978). For instance, Levine *et al* (1979b) have demonstrated that in CO_2 rich atmospheres lightning is an important source of CO. Even today, volcanic emissions contain significant, but variable, amounts of this gas (Brancazio and Cameron 1964). Carbon monoxide is a significant secondary component of the martian atmosphere (see e.g. Walker 1977). The main carbon monoxide absorption feature is a vibration–rotation band at 4.65 μm, with a weaker band at 2.40 μm. The change in terrestrial surface temperature has been calculated allowing for variations in carbon monoxide partial pressure from 0.001 to 1 mbar with various values of the total atmospheric pressure. From table 4.3 it can be seen that even major changes in carbon monoxide content have little effect on surface temperature.

144

So far as minor atmospheric carbon compounds are concerned, it seems, therefore, that variations in the reduced form (CH_4) are more important for surface temperature fluctuations than are variations in the partially oxidised form (CO).

Ammonia. Ammonia has been considered (e.g. Sagan and Mullen 1972) as a candidate for enhancement of early surface temperatures because of its considerable infrared absorption. It has been argued that ammonia must be rapidly removed from the terrestrial environment (Ferris and Nicodem 1972, Kuhn and Atreya 1979) for a number of reasons. (i) Ammonia dissolves readily in water. NH_3 is lost rapidly by 'rainout' (e.g. Levine *et al* 1980a). (These authors calculate that in the present atmosphere the lifetime of NH_3 against rainout is about 10 days.) However, this removal mechanism will be balanced by upward diffusion from the ocean surface once the oceans become saturated with ammonium ions (Kasting 1982). (ii) Photolysis of NH_3 rapidly removes it from the atmosphere. This may be important since many models of chemical evolution require significant concentrations of NH_3 and NH_4^+ in solution for the synthesis of amino acids and other biochemicals. For example, it is estimated that concentrations greater than 10^{-3} M of NH_4^+ are needed to permit the existence, at equilibrium, of equal concentrations of amino acids and the corresponding hydroxy acids at pH 8 (Bada and Miller 1968).

There are, however, a few mechanisms for the synthesis of NH_3 (Kasting 1982). NH_3 is produced *in situ* through the decomposition of HCN in aqueous solution. Thus water bodies could be a source of atmospheric NH_3 but this source is critically dependent upon other elements in the aqueous environment (e.g. metal ions). Recent work on the direct photolytic reduction of N_2 on natural, semiconductor catalysts (titanium dioxide) suggests a massive process which would have operated to increase localised steady-state concentrations of NH_3 near ground level and, more importantly, in associated sources of ground water (Henderson-Sellers and Schwartz 1980). It should be noted that Wigley and Brimblecombe (1981) suggest that the required level of atmospheric

145

and/or environmental NH_3 is very much lower than had previously been discussed. Recently Kasting (1982) has re-examined the sources and sinks of NH_3 in a pre-biological atmosphere. The results of his $1D$ photochemical model suggest that the abiotic source described by Henderson-Sellers and Schwartz (1980) is large enough to maintain mixing ratios of the order of 10^{-8} but ratios large enough to produce significant warming ($\simeq 10^{-5}$ or more) are unlikely due to the continuing destruction of NH_3.

Henderson-Sellers and Schwartz (1980) have calculated the enhanced surface temperature in the evolutionary environment using an approach similar to that described in §3.2 and the appendix. The major regions of NH_3 band absorption are 2.8–3.2; 5.5–7.2; 8.0–15.0 and longwave of 35.0 μm. Table 4.4 lists the 'greenhouse increments' for the addition of 10^{13} kg of ammonia to an evolving atmosphere of CO_2 with the present-day level of water vapour. From this model calculation it seems possible that 10^{13} kg of NH_3 in the evolving atmosphere of the Earth may produce surface temperatures too high for the subsequent evolution that is known to have occurred. However, it is important to note that the calculations reported in table 4.4 use CO_2 levels from Hart (1978) which may be up to an order of magnitude too high for the earliest epochs.

The question of the possibility of a climatic catastrophe has been discussed (Hart 1978, Schneider and Thompson 1980, see also chapter 6). The geological data (Miller and Orgel 1974, Windley 1976) do not support any history which has

Table 4.4. The greenhouse effect caused by ammonia in the early Earth's atmosphere for the addition of 10^{13} kg NH_3. Temperatures from Henderson-Sellers and Schwartz (1980). Temperatures in brackets are the recomputed ΔT values when the convective adjustment is made (after Henderson-Sellers 1982).

Time (10⁹ years BP)	pCO₂ (mbar)	10^{13} kg NH_3 ΔT (K)
4.25	310	17.2 (4.2)
3.5	70	15.1 (3.8)
2.5	18	13.7 (3.6)

either very high early surface temperatures or very large mixing ratios of ammonia. However Henderson-Sellers and Schwartz (1980) failed to include any convective adjustments in the surface temperature calculations. Surface temperatures increased by up to 17 K must surely result in considerable tropospheric reorganisation. Specifically the values of T_{top} and T_s have been recomputed as described in §4.2.6 and ΔT values are given in table 4.4. These recomputations fail to include the cloud albedo feedback which would also, presumably, result (see the discussion in §2.3.2). Despite these omissions it appears that the conclusions of Henderson-Sellers and Schwartz (1980), namely that 10^{13} kg of NH_3 is an upper limit to the atmospheric level of ammonia compatible with known geological data, remain valid. This abiological source of a significant reduced trace gas illustrates the extremely complex nature of atmospheric/environment reactions. Two other trace gases are significant for life if not directly for the climate. Oxygen and hydrogen cannot be discussed separately from one another. The discussion below draws upon the much more complete evaluations of e.g. Walker (1977) and Kasting and Walker (1981).

Hydrogen and Oxygen. The mixing ratio of hydrogen in the evolutionary environmental atmosphere is now generally believed to be very small ($\geq 10^{-5}$). However, its presence may be of considerable importance for abiological synthesis of simple organics and further it may prove an important source of energy in early metabolic evolution (Walker 1978a). Free hydrogen, even as a very minor trace constituent, will be highly significant for many chemical and photochemical reactions and will also control the level and evolution of oxygen in the atmosphere. The assertion that photolysis of water vapour provides a source of free oxygen in the pre-biological atmosphere is recurrent in the literature (e.g. Walker 1978b). Tropospheric photolysis is usually rapidly followed by recombination reforming H_2O. Only escape to space of hydrogen can result in a net atmospheric source of oxygen. The geological data demand a reducing atmosphere prior to about 2.0×10^9 years ago. If banded iron formations result from oxidation by atmospheric O_2 then these formations indicate the need for a considerable supply of free oxygen. Holland's data suggest (Walker 1978b) that a single banded iron formation would require a supply of O_2 16 times larger than the present-day source provided by the photochemical destruction of water vapour. It is important to note that there are alternative methods by which the banded irons could have been formed. For instance (Randall 1982) has shown that irradiation of Fe^{2+} solution will deposit Fe^{3+} salts and banding could be a subsequent process brought about by solar fluctuations. The appearance of the redbeds dates from about 2×10^9 years BP (Cloud 1968, 1976). The occurrence of this distinctive formation may indicate free atmospheric oxygen but the level could be as low as 10^{-8}–10^{-7} of the present atmospheric level (PAL) (Kasting and Walker 1981). The banded iron formations probably require reducing waters if the iron is to be transported in solution suggesting a complex but quasi-equilibrium situation amongst the various processes controlling environmental levels of oxygen and hydrogen. For instance the O_2 contained in these deposits could have been produced locally by oxygenic photosynthesis. Walker (1978b) believes that increased tropopause temperatures are unlikely to have been responsible for enhanced levels of O_2. However, the considerable greenhouse increment caused by, for instance, the addition of abiologically fixed NH_3 as described above could result in greatly enhanced tropospheric temperatures (see table 4.4). Figure 4.4 illustrates the major present-day sources and fluxes of atmospheric oxygen. Walker (1977) suggests that the mixing ratio of free oxygen is only a weak function of the oxidation state of surface minerals in the weathering process since the primary reaction is with reduced volcanic gases, primarily hydrogen. If abiological processes such as those simulated in laboratory experiments were global in extent (Sanchez et al 1966, 1967, Toupance et al 1975, 1978) and persistent for a significant time then these might in themselves compose a non-negligible source of free hydrogen in the evolutionary environment. The consideration of possible levels of these trace constituents in the atmosphere is as dependent upon environmental feedbacks as is the level of, for instance, CO_2 discussed in chapter 3.

148

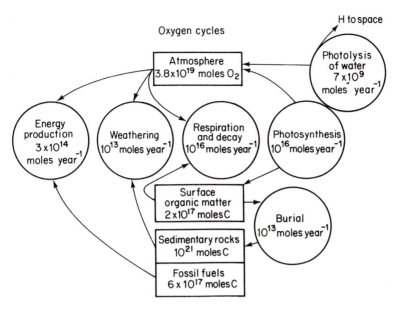

Figure 4.4. Major reservoirs (rectangular boxes) and pathways of oxygen within the Earth's atmosphere–hydrosphere–biosphere–lithosphere system (after Walker 1977).

The atmosphere is of importance in (i) the provision of hospitable environmental conditions for the genesis and evolution of life; (ii) the provision of source areas of e.g. abiologically produced NH_3; (iii) as a secondary factor controlling gaseous levels through weathering; and (iv) direct weather effects, for instance, the solution of atmospheric gases as a result of, or enhanced by, precipitation processes and the direct formation of trace gases by lightning. Chameides *et al* (1979) discuss the production of a considerable range of gaseous constituents by lightning in the atmospheres of Venus and Mars. Extrapolation of their results to the evolutionary environment indicates that lightning (which must have existed) would produce CO, O_2, NO and O (see also Levine *et al* 1979b). A further, probably minor, effect is the formation of trace compounds in the upper atmosphere directly as a result of meteoritic impact, see e.g. Henderson-Sellers (1977) and earlier in this chapter.

The seven characteristics of the evolutionary environment are vital to any discussion of the evolution of life. It is evident that the evolutionary environment resembles the

149

present-day situation on Earth in many ways. For instance, similar figures for the tropospheric lapse rate, average global surface temperature and zonal energy transport indicate a climatic regime not too dissimilar from that which is experienced today. The definition of the atmospheric and hydrological characteristics of the planetary environment, within which it is believed life originated, separates the case of the Earth from general models of planetary atmospheric evolution. The latter part of this chapter will be devoted to an examination of the interactions between an existing but still evolving biosphere and the atmosphere. Other short-period changes are discussed in chapter 5.

4.3. Evolution of the Earth's Climate, Hydrosphere and Biosphere

Possibly the most important environmental upheavals on the Earth have been a direct consequence of the evolution of life. For instance the release of free oxygen to the previously neutral or mildly reducing environment must have caused upheavals in geophysical processes. Changes in weathering and sedimentation (see figure 5.1) are illustrated by the peak in production of banded iron formations and by the transition between the formation of banded iron and redbeds. Climatological effects of a net oxygen source are probably negligible but the almost simultaneous development of a substantial ozone screen in the stratosphere must have perturbed both surface temperatures and environmental lapse rates (see e.g. Levine and Boughner 1979). Figure 4.5 illustrates the likely changes in these parameters as calculated by a $1D$ radiative–convective climate model. For a transition from zero $O_2 + O_3$ to the present-day levels the surface temperature increases by only approximately 2.3 K but the stratospheric profile is totally unchanged. This surface temperature change is much less than the temperature changes caused by increased levels of ammonia discussed in §4.2.7 above. NB This perturbation also could affect cloud cover through changing the amount and nature of convective activity (§2.3.2).

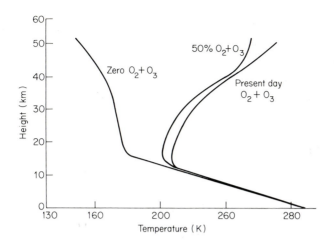

Figure 4.5. Vertical temperature profiles calculated with a 1D radiative–convective model for three different O_2+O_3 concentrations: (i) present day, (ii) 50% of current O_2+O_3 value, (iii) zero O_2+O_3. The model calculations assumed fixed relative humidity and fixed cloud temperature (for description of the model see Hansen *et al* 1981) (after Henderson-Sellers 1982).

The level of most of the trace gases in the evolutionary environment must have suffered considerable disruption during the origin and evolution of life itself. Chemical upheavals within the troposphere may require detailed examination, unless life evolved extremely quickly (say less than 10^6 years) or the organic precursors already existed on the surface of the Earth as a result of final planetary accretion. It should be noted here that at least two separate lines of evidence (i) carbon isotope chemistry (Schidlowski 1980a) and (ii) the presence of a net oxygen source in the environment (Walker 1978b) could point to the appearance of life on Earth prior to 3.8×10^9 years ago. It is possible therefore that life may have arisen quickly.

The classical Urey–Miller experiments produce considerable quantities of free hydrogen as well as the required organic materials (from a mixture of gases now deemed unlikely within the terrestrial environment). The results of Toupance *et al* 1975, 1978) in which numerous, more plausible, gaseous mixtures have been irradiated with ultraviolet radiation do however indicate that considerable disruption of the environment would have occurred. These may include a change in the ratio of oxidised to reduced forms of various compounds which could modify the environment by, for instance, changing the infrared absorption characteristics of the atmosphere (see §4.2.7 above) and the evolution of free atmospheric oxygen and ozone.

Reactions of interest to origin-of-life biochemists appear to be functionally dependent upon the level of H_2 within the environment. At its simplest level Toupance and his co-workers find that the production of the molecules believed to be of importance for the evolution of life is inhibited in an NH_3/CH_4 mixture if H_2 is retained and the ambient level allowed to increase. Within a CO_2/N_2 mixture retention of H_2 aids the evolutionary process. This work substantiates that of Abelson (1966) in which production of amino acids from a mixture of CO_2, CO, N_2, H_2O and between 1 and 10% H_2 was achieved by ultraviolet irradiation. Unfortunately environmental levels of H_2 of this order are very difficult (if not impossible) to sustain within the constraints of the characteristics of the evolutionary environment defined above. Walker (1978b) calculates that a volcanic hydrogen source 1000 times larger than that of the present day would be required to give hydrogen mixing ratios in excess of 1%— Walker's (1978b) calculation neglects the sink of oxygen provided by the weathering of reduced surface minerals. This omission is not important if the atmosphere is at least mildly reducing. In any case it seems most unlikely that levels of H_2 ever exceeded about 1%.

It is interesting to note that the evolution of oxygenic photosynthesis does not necessarily coincide with the transition to an oxidising atmosphere (Berkner and Marshall 1965, Walker 1977). Kasting (1979) suggests that precisely the opposite occurs: the initial result of the advent of photosynthesis is a more heavily reducing atmosphere. The sequence is

$$H_2O + CO_2 \xrightarrow{\text{photosynthesis}} CH_2O + O_2$$

$$O_2 \xrightarrow{\text{Fe}^{2+}} Fe_2O_2 \text{ (solid)} \qquad (4.1)$$

$$2CH_2O \xrightarrow{\text{fermentation}} CH_4 + CO_2.$$

By splitting the water molecule and tapping the reducing power of ferrous iron, photosynthesis could have provided a significant source of atmospheric *methane*, or some other reduced compound like H_2. Oxygen levels close enough to the present atmospheric level (PAL) to behave in a chemically similar way (but see also Kasting and Walker 1981) were certainly established by 2.0×10^9 years ago although the presence of redbeds may require little free oxygen. The evolution of the ozone layer in the stratosphere would have been a direct result of increasing levels of tropospheric O_2 (Kasting and Donahue (1980) and see chapter 2 and figures 4.5 and 4.6).

Levine *et al* (1980b) have made a detailed study of the evolution of atmospheric ozone as a function of increasing oxygen levels. They find that surface and tropospheric levels of O_3 and total column amount reach a maximum for an atmospheric O_2 level of 10^{-1} PAL. The calculated profiles of O_3 are shown in figure 4.6. The increased O_3 production in the lower atmosphere seems to be the result of larger numbers of third bodies to take part in the formation process without a corresponding increase in destruction rates. Levine

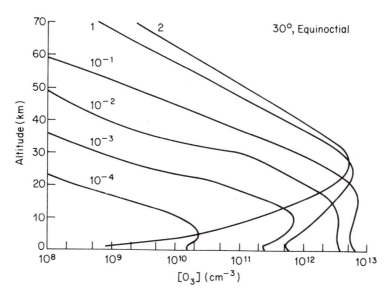

Figure 4.6. The vertical distribution of O_3 as a function of atmospheric O_2 level (after Levine *et al* 1980b).

153

et al (1980b) calculate that when 10^{-1} PAL was achieved surface temperatures were likely to have been around 4.5 K higher. However, it is important to stress that their computations do not include many important tropospheric processes such as a surface sink, 'rainout' and CH_4 oxidation. Levine (1980) has used these computed O_3 profiles to calculate the levels of ultraviolet radiation penetrating to the surface of the Earth in various epochs.

It has been suggested that feedbacks between atmospheric oxygen levels and the oceans through O_2 solution could have affected tropospheric partial pressures. Henderson-Sellers and Henderson-Sellers (1978, 1980) suggest that the variation in pycnocline/thermocline (the region of rapid salinity/temperature change) depth may have controlled the atmospheric level in a way that fed back into the developing biosphere. They find that the early stages of diversification of land biota may have been at least partially encouraged by oxygen concentrations in the upper levels of oceans and lakes.

The authors argue that during periods of stratification (whether permanent or not) aerobic conditions are confined to the water above the pycnocline. When mixing temporarily occurs, some oxygen is conveyed to the deeper regions of the water body but is unlikely to be able to equal the concentrations in the upper layer until after the advent of oxygenic photosynthetic and respiring forms of life and the establishment of the present-day aquatic ecosystem (i.e. eukaryotic bacteria†). In addition, a certain depth of water ($\sim$10 m for distilled H_2O but much less for oceanic waters) is needed (Margulis *et al* 1976) to provide a protective barrier for all multicellular organisms against ultraviolet radiation. The changes in pycnocline depth (figure 4.7) may have influenced dissolved oxygen concentrations and hence the evolution and diversification of life.

Henderson-Sellers and Henderson-Sellers (1980) believe that the large depth of the mean pycnocline between 4.5×10^9 and 3.5×10^9 years ago (see figure 4.7 and Sutton and Windley 1974) afforded protection to the evolving organ-

† It must be noted that these bacteria will also be limited by nutrient sources and temperature fluctuations.

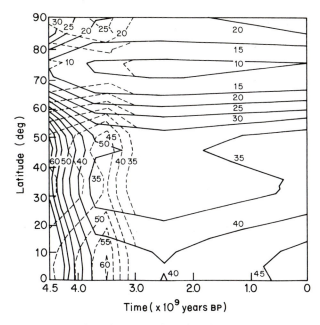

Figure 4.7. Predicted pycnocline depths as a function of latitude and time for the 4.5×10^9 years of the Earth's evolution. Full lines are without lunar capture effects and the broken lines include these (after Henderson-Sellers and Henderson-Sellers 1980).

isms. Over the period following, until about 2×10^9 years ago, the upper mixed layer decreased in thickness. Photosynthetic plants in this layer would have continued to produce oxygen within the water body. The time of the minimum pycnocline depth corresponds to the maximum rate of transfer of oxygen to the atmosphere, thus augmenting atmospheric oxygen and hence ozone. Such conditions of high dissolved oxygen concentrations would have been ideal for the rapid diversification of life. Since no fossil evidence for respiring metazoa at this time exists it is suggested that where the pycnocline depth was small enough to permit high oxygen levels, the water depth was inadequate to afford protection against ultraviolet radiation. Calcareous skeletal features seem to require dissolved oxygen concentrations corresponding to an atmospheric oxygen level of at least 10% of the present-day value (Goldring 1972); this value was attained 0.8×10^9 years BP and land animals were in existence some $0.1–0.2 \times 10^9$ years later. Thus multicellular life

155

may have evolved at some time during the period 2×10^9 to 0.8×10^9 years ago (Flint 1973).

Interactions between the hydrosphere and atmosphere for the evolving biosphere are important. In chapter 6 further arguments for the necessity of the existence of a planetary hydrosphere for the support of the evolution of life are presented. The effects of biospheric evolution can be viewed as a perturbing force acting upon a naturally stable hydrosphere–atmosphere system. For instance, two direct results of the evolution of life and particularly the evolution of green plant photosynthesis are the increase of free oxygen in the atmosphere (figures 4.4 and 4.5) and the subsequent development of an ozone shield in the stratosphere (figure 4.6), both of which constitute traumatic changes to the environment. In particular, two facets of biological perturbations may be worthy of brief mention. The first relates to the possible perturbations to the evolutionary environment caused by pre-photosynthetic life. For instance, Walker (1978a) has suggested that the first organisms removed the organic molecules which had been abiologically synthesised whilst later evolutionary steps led to the direct exploitation of the chemical energy available in the environment. For instance methanogenic bacteria† use H_2O as the hydrogen source with which to reduce CO_2. As global scale evolution progressed organisms themselves may have dominated or at least considerably perturbed their environment.

Suggestive coincidence occurs between the period of rapid oxygen build-up in the atmosphere and the first evidence for large-scale glacial events. The geological record shows glacial activity between 2.5 and 2.0×10^9 years ago (see figure 5.1). It is possible to argue that biospherically initiated changes in the chemical composition of the atmosphere (either a reduction in the carbon dioxide level, or, more likely, the rapid removal of traces of reduced constituents following the evolution of free atmospheric oxygen) were the trigger for these glacial events.

The second aspect of biological perturbation relates to the Gaia hypothesis (see e.g. Lovelock 1979). Lovelock believes

† Many methanogens can also cleave acetate and reduce the resulting methyl group to CH_4 regenerating CO_2 or sometimes HCO_3.

156

that the presence of life perturbs environmental systems in such a way as to control and even dampen instabilities. He suggests that extensive surface albedo changes due to the biosphere and production within the biosphere of trace gases which absorb in the infrared, e.g. methane, ammonia, are symptoms or examples of the cybernetic control mechanism exercised by Gaia in the current terrestrial environment. Figure 4.8 forms an interesting comparison with figure 2.13 since it shows the smoothed upward trend of CO_2 in the post-industrial atmosphere. Small oscillations previously indistinguishable could be interpreted by Lovelock as the partial success of a Gaia control mechanism, apparently only just invoked. An alternative (and possibly more satisfactory) explanation depends simply on the natural feedback effects within the complete environmental systems and specifically (in this instance) on the sea surface temperature anomalies which enhance the CO_2 uptake by the oceanic biomass. Thus

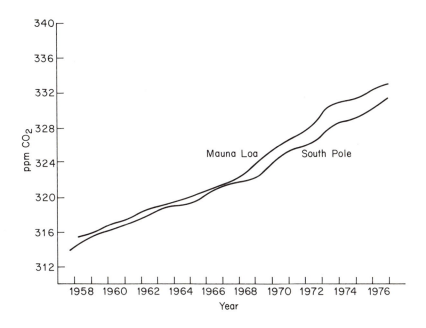

Figure 4.8. Seasonally adjusted trends in atmospheric CO_2 on the Earth as observed at Mauna Loa and at the South Pole (after Bolin *et al* 1979). When biospheric effects are smoothed (cf figure 2.13) the weaker signal due to the hydrosphere–biosphere southern oscillation effects are seen.

it is possible that the hydrosphere is the *dominant* factor in environmental feedback processes, even those operating on extremely short time scales of a few years (see chapter 6). Certainly many of the biospheric feedback processes cited by Lovelock (1979) as important controlling factors in the environment can operate only within or assisted by the hydrosphere. The response of planetary atmospheric systems to short-term external and internal perturbations will be considered in chapter 5.

The likelihood of hospitable conditions throughout the evolutionary environment has been underlined here. There are a number of reasons why some biochemists (e.g. Lahav *et al* 1978, Schwartz 1981) favour the terrestrial environment as the site of the origin as well as the evolution of life. There is a wide range of organic molecules which can be synthesised in a gaseous environment and also in space (e.g. Chang *et al* 1981). However, despite the complexity of these compounds, it remains uncertain whether further synthesis is likely or even possible except within the cells themselves without the presence of mineral surfaces. A solid surface can serve as an organisational base for chain formation (e.g. Matheja and Degens 1971). Whilst chemical evolution is thermodynamically possible within the atmosphere–hydrosphere system it is the interactions with parts of the lithosphere which render the processes productive on a global scale.

The evolutionary history of the Earth's atmosphere seems to have been dominated by the fact that the global mean temperature has remained fairly constant throughout geological time (figure 2.2 and the appendix). This fact has permitted the existence of a global hydrosphere and established and allowed the persistence of hospitable conditions for life. The question of the uniqueness of the Earth's atmospheric evolution (Henderson-Sellers 1981) (see table 4.2) will be considered again in chapters 5 and 6.

5. Planetary Climatology: Shorter Timescale Atmospheric Changes

This work is an investigation of the evolution of planetary atmospheres and thus most of the preceding discussions have been for time periods longer than about 10^8 years. Over these time periods many fluctuations likely to perturb the atmosphere have to be ignored. Dismissal of shorter time-scale factors may not be warranted in spite of the predominant interest only in geological time periods. This is because short-period impulses could possibly cause internal feed-backs within the planetary climate system, which in turn could cause fundamental long-term changes. For instance, an extreme example would be a collision between a planet and a comet which would disturb and possibly totally disrupt the atmosphere (see e.g. Lazcano-Araujo and Oro 1981). A critical question relating to the problem of the significance and importance of time-scales is evident in the history of the evolution of the atmosphere of Venus (chapter 3). It is not known how long surface temperatures must remain above the critical threshold of around 350 K to sustain the initial stages of the 'runaway' process.

This chapter focuses upon shorter-period changes which may be of importance for the long-term evolutionary history of the planet. An excellent review of planetary climatic change *per se* is given by Pollack (1979). It is only interesting to study short-term or 'climatic' changes in planet–atmosphere systems that are fairly but not completely stable. This chapter is mostly concerned with the planets considered in

159

§3.4 particularly Mars and the Earth. The complexity of interactions between shorter-period (and periodic) excursions of climate and the planetary evolutionary history are very difficult to untangle. For instance for the Earth, for which there is the most data, neither short- nor long-term forcing effects seem to have resulted in a large shift in the mean planetary conditions (see chapter 4). For this reason most timescales of possible climatic change mechanisms are mentioned here for the Earth. In the case of Mars, however, perturbations which might be classed as both short-period and secondary for the Earth (e.g. Milankovitch variations in the solar radiation received) may turn out to be the cause of large-scale atmospheric loss (§5.2). It is likely that the same mechanism is causing very different effects in these two planetary regimes and must therefore be considered somewhat differently.

5.1. Climatic Forcing Effects on the Planets

Pollack (1979) suggests that the major factors affecting climatic regimes on the terrestrial planets can be subdivided by their spatial location. Amalgamating two of his original categories we have:

  (i)   solar radiation at the top of the atmosphere;
  (ii) the atmosphere;
  (iii) land surface and volatile deposits;
  (iv) the planetary interior.

Each of these 'climatic' factors can be affected further by perturbations which fall into the more traditional classification of either (*a*) external or astronomical or (*b*) internal or geo(planetary)physical. External and internal variations in each of the four categories above will be discussed and also a fifth subdivision of 'diverse' will be included. An attempt is made to specify time periods and the temporal extent of activity but no consideration of feedback effects is made although these are acknowledged to be fundamental to the system.

5.1.1. *Solar Radiation at the Top of the Atmosphere*

Variations in the incident solar flux external to the planetary system are a function of either (or both) a change in the total

160

luminosity of the Sun (or the spectral distribution of the solar radiation) or perturbations in the orbital parameters of the planet itself. Clearly changes within the atmosphere–surface system may significantly affect the planetary albedo (Henderson-Sellers and Meadows 1979a) and hence internal mechanisms may be as strong or stronger perturbing influences on the absorbed solar radiation than external factors.

It is generally believed that there has been a long-term increase in the solar luminosity of between 25 and 45% over the lifetime of the solar system (e.g. Haselgrove and Hoyle 1959, Newman and Rood 1977, see also chapter 1). The timescale of this stellar evolutionary change is therefore of the order of 10^9 years. There are also numerous suggestions in the literature that solar output may vary on shorter time periods, either as a function of internal stellar dynamics (Christensen-Dalsgaard and Gough 1976, 1981) or photospheric activity, e.g. the sunspot cycle (Schneider and Mass 1975). Influences external to the solar system itself may also affect the amount and spectral nature of flux at the planet (McCrea 1975). Clark *et al* (1977) have discussed the effects of the passage of the solar system through dust and gas clouds in the galaxy. This process seems likely to have a characteristic time of the order of 10^8 years, whilst the short-term stellar fluctuations described above may range in period from 10^7 years down to as little as seconds.

Berger (1979) has made an excellent review of the perturbations to orbital and axial characteristics of the terrestrial planets (and especially the Earth) through what is known as the Milankovitch mechanism. Here the periodicities are approximately 10^5, 4×10^4, 2.3×10^4 and 1.9×10^4 years for the Earth, but differ considerably for the other terrestrial planets. For instance both the relative magnitudes and periodicities of the eccentricity, obliquity and precession changes differ between the Earth and Mars (Berger 1980).

The effectiveness of the solar signal as a perturbing mechanism to the planetary climate system is also a function of the planetary rotation rate and the optical characteristics and overall mass of the atmosphere. The importance of these parameters has been discussed by Henderson-Sellers and Meadows (1975, 1976) in terms of the relative magnitude of the planetary rotation rate, τ_r and the thermal relaxation time of the atmosphere τ_h (Goody 1975, Golitsyn 1979) (table 3.7 and §2.1) which determine the value of the planetary flux factor (see chapter 3).

Despite the stated aim of limiting the discussion of feedback effects, it can already be seen that perturbations in and response to as apparently an external signal as solar luminosity are intimately and undeniably linked. Furthermore, the considerable difficulty in establishing time periods is evident. The solar signal is likely to have evolved over periods of the order of 10^9 years being modified, possibly in a cyclic way, on periods of about 10^8 years; celestial configurations cause both signal and feedback changes with periods of about 10^2 (or possibly less) years to 10^6 years and there may also have been signal and response perturbations with periods less than 10^2 years.

5.1.2. Atmospheres

The atmosphere of the Earth (and probably, of all the terrestrial planets massive enough to retain an atmosphere) is of secondary origin, see chapters 1, 3 and 4. This has been generally accepted since the comprehensive assessment of possible sources for the 'excess volatiles' by Rubey in 1951 (table 2.5a). It has also been suggested (Benlow and Meadows 1977, Lazcano-Araujo and Oro 1981, Levine et al 1981) that some constituents of the atmospheres of the terrestrial planets could be due to external sources. Accretion of atomic particles due to the solar wind combined with the impacting of dust and larger amounts of carbonaceous meteoritic material early in the history of the planets are discussed in chapter 2. Whichever of these processes dominated the initial stages of atmospheric formation around the terrestrial planets, it seems reasonable to assume† that the gases involved were similar for all these planets and that the differences apparent in the present-day atmospheres are due mainly to differences in mass and distance from the Sun of the planet itself (see chapter 3).

Degassing processes are evident on the Earth and Io today

† This is a first-order approximation only, see the discussion of figure 3.5.

in volcanic events and evidence has been cited in chapter 3 which indicates volcanic degassing on Mars and Venus in earlier epochs. Surface features typical of both crustal/volcanic events and meteoritic impacts have been identified on all the terrestrial planets and some of the 'icy' satellites (§§3.3 and 3.4). The juvenile gases detected in volcanic events today on the Earth are typically H_2O, CO_2, CO, N_2, SO_2 and HCl and a few other trace gases. These, probably in the given order by volume, have been degassed throughout the history of the Earth. However the differentiation of the planet may have influenced not only the degassing rate, but possibly the chemical state of the gases evolved (Walker 1977 and chapter 4).

The composition of the atmosphere of a planet at any point in its history depends upon the net fluxes at the top of the atmosphere and at the surface, the surface temperature, planetary mass and distance from the Sun (chapters 2 and 3). Whilst the climatic modification due to atmospheric gases is almost entirely a result of their contribution (either directly or indirectly, e.g. through pressure broadening effects) to the 'greenhouse' effect, atmospheric aerosols can cause either net warming or cooling depending upon the relative magnitude of interactions with both the visible and infrared radiation streams. An increase of aerosol content in the troposphere generally increases infrared opacity and hence enhances the 'greenhouse' effect. The dramatic effects of the increase in optical depth during the global dust storm on Mars in 1971 is commented upon by Levine *et al* (1977). They calculate that the mean annual daily insolation at the poles decreased by more than a factor of 100 as τ increased from 0.1 to 2.0. Discussion of wind-blown dust on Venus is found in Sagan (1975a,b). It is important to note that removal processes are at their most efficient within the troposphere of a terrestrial planetary atmosphere. Thus without replenishment, it may be anticipated that greenhouse enhancement by aerosol increase is a fairly short-lived effect (say less than 1 year). Weather processes (and on the Earth also anthropogenic effects) clearly provide replenishment of these aerosols. Calculation of the effect of atmospheric aerosols upon the planetary albedo through scattering of visible radiation turns out to be a very complex process dependent upon the albedo

163

of the underlying atmosphere and the surface itself (Chylek and Coakley 1974). Also the vertical situation of these aerosol layers is critical both in terms of their climatic effect and the likely period of residence (Hansen *et al* 1980). For instance, the cytherean upper troposphere consists of a dense layer of droplets of sulphuric acid which is fundamental in determining the planetary albedo. Global scale dust storms on Mars seem to be both cause and response to minor cyclic climatic variations (Murray *et al* 1973, French and Gierasch 1979, Spitzer 1980). On the Earth, the energy of certain volcanic eruptions is great enough to eject particulates into the stratosphere and in this case the climate signal is increased both because of the comparative weakness of removal mechanisms in the stratosphere and the larger absolute signal generated (Lamb 1970, Hansen *et al* 1980). Aerosol perturbations to the planetary albedo may, it appears, be either very short-period changes or may persist for between 1–10 years; whereas the timescales over which changes in atmospheric gaseous constituents may be a perturbing influence upon planetary climate must be dependent upon the mechanisms controlling degassing, loss to space, chemical reaction and condensation of volatiles which in turn may be a function of the greenhouse effect itself.

Geological evidence on the Earth (Schidlowski *et al* 1979) now indicates that surface pressures must have been of the order of at least 500 mbar by 3.5×10^9 years ago (Henderson-Sellers 1981). The effects of increasing partial and total pressures of important greenhouse gases have been investigated by Henderson-Sellers and Meadows (1979b) and Owen *et al* (1979). Time periods are therefore of the order of 10^9 years, since whilst charges are restricted to the earliest 1×10^9 years for the Earth, degassing may have been a much slower process on e.g. Mars. All other fluctuations in gaseous levels and constituents are under the control of feedback processes.

5.1.3. Land Surface and Volatiles

The planetary surface configuration (i.e. large-scale topographic features) and the surface position of volatile deposits are dependent upon one another. For instance the time required

164

to form and flood oceanic basins is still unclear. The latest geological data suggest that for the Earth crustal differentiation (into basaltic 'continents' 'floating' on a sialic lower layer) had taken place on the Earth by 4.1×10^9 years ago (Schopf 1980, Moorbath 1982). Thus ocean formation depends upon the rapidity of condensation of outgassed water vapour. In any planetary evolutionary history the system albedo will initially be very low (~0.07, the current lunar value). As atmospheric mass is acquired, surface deposition of volatiles and later, atmospheric condensation or freezing of volatiles will increase the albedo. The rate and nature of this albedo increase is unclear. Liquid phase volatiles will be, as described above, constrained by the general surface topography, whilst deposition by freezing will be as strong a function of local temperature (i.e. occurring around polar regions and possibly in topographic areas similar to 'frost hollows' that occur on the Earth and in martian craters (Thomas *et al* 1979)). System albedo changes must therefore be strongly constrained by feedback processes within even the earliest stages of evolutionary history (see chapter 3) and are likely to have temporal signatures ranging from very much less than 1 year through to 10^9 years.

Topographic features on any terrestrial planet must be the result of both internal heat transfer and rock cycling (see chapters 3 and 4) and the result of impacts of meteoritic-type material (chapter 2 and §3.4). The size of the debris may range from the size of some of the smaller asteroids (or planetesimals—Wetherill (1980) shows planetesimals of radius around 800 km) through to dust of less than 1 μm in diameter. Evidence for considerable and extended periods of bombardment has now been derived from studies of the Moon, Mars, Mercury and the 'icy' satellites. It is interesting to note that many of these bodies and indeed the Earth itself exhibit an, as yet unexplained, marked hemispheric asymmetry. Hemispheric asymmetry is also a feature of some satellites which are tidally locked to the parent. In this case the asymmetry could result from long-term seasonality (e.g. Titan's polar hood) or as a result of sweeping or deposition favouring one side of the moon (e.g. Iapetus). Analysis of impact features on the Earth is hindered by the continuing

165

geological and geomorphological reprocessing of the surface. Preliminary studies by Grieve and Dence (1979) allow tabulation of impact events and an estimation of the likely interval between large impacting objects ($\sim 10^8$ years), see McCrea (1981). These impacts can be viewed as the protracted 'tail' of a process which has been continuing since the late heavy bombardment around 4.1×10^9 years ago. The early details of this process are unknown but more than one 'subpopulation' has been identified (Smith *et al* 1981) and there is some evidence for post 4×10^9 year impacts on various bodies—for instance meteoritic impact on the Earth. It seems reasonable to conclude that on a system-wide basis the collision rate is still non-zero. Thus viewing individual events as catastrophic may be unwise.

Surface albedo variations have often been invoked as important mechanisms of climatic change on the Earth. The major themes in the literature are reviewed by Henderson-Sellers and Hughes (1982). Charney (1975) suggests a bio-geophysical feedback mechanism as the cause of extending drought regions in the Sahel. Increased albedo caused by overgrazing results in a net radiative loss which produces general subsidence and drying over the area, thereby inhibiting or reducing the convection necessary for rain. This theory has been tested in general circulation climate models by a number of groups (e.g. Charney 1975, Walker and Rowntree 1977) and disputed by Ripley (1976). The specific studies undertaken by Otterman (1975, 1977) into the albedo contrast between the Sinai and Negev regions along a dividing fence (a result of overgrazing in Sinai by the Bedouin) suggests that the higher albedos in Sinai resulted in lower surface temperatures, producing a decreased lifting of air and hence reduced cloud and rainfall. Micro- and meso-scale studies of the feedback effects of local albedo changes have also been considered by Berkofsky (1976) studying albedo variations in the Negev desert. Similar mechanisms have been suggested as the 'trigger' for martian dust storms (Leovy and Zwek 1979, Cutts *et al* 1979).

Planet-wide dust storms are common on Mars as it approaches perihelion. Such a storm obscured the surface for the early part of the Mariner 9 mission. Smaller storms were also observed by the Viking orbiter cameras and their effects detected by the meteorology packages on the landers (Spitzer 1980). The global storms observed by Viking are both believed to have started in the Thaumasia–Solis Planum region. Localised storms are often generated near the edge of the retreating south polar cap and in the irregular terrain on the SE slopes of Tharsis (French and Gierasch 1979).

Local or hemispheric cooling may trigger glaciation. The climate sensitivity testing of Lian and Cess (1977) presents a reappraisal of the ice albedo feedback mechanism. They suggest that the ice albedo feedback (§5.2) amplifies global climate sensitivity by 25%. Cess's (1978) climatic sensitivity model is also used to consider the effect of vegetation modification of the Earth's surface albedo at 18 000 years BP. This albedo change results in a doubling of the global climatic sensitivity to factors forcing climatic change. Bray (1979) discusses the surface albedo increase in the northern hemisphere from ash deposited by explosive Pleistocene eruptions, the cooling resulting from this helping to trigger glaciation.

The early work of Ewing and Donn (1956) pertaining to continental drift and climatic change has been developed more recently by Henderson-Sellers and Meadows (1975). The likelihood of extensive glaciation occurring should be greater when a higher proportion of the land mass is near the poles—resulting from the changed surface albedo. Following a similar theme, Cogley (1979) suggested, on the contrary, that a concentration of land in the tropics, as a result of continental drift, provides a necessary condition for glaciation. His work used improved ocean albedo values but omitted the effects of cloud cover. More recently the effect of land–sea distribution has led Barron *et al* (1980) to the conclusion that surface albedo variations between 0.18×10^9 years BP and the present day are a strong function of sea and lake level fluctuations.

Rampino *et al* (1979) have recently put forward an interesting hypothesis which, whilst once again linking tectonic activity and glaciations, argues that, contrary to the general supposition, it is land glaciation which may trigger volcanic activity. This type of second-order environment–climate feedback is examined in greater detail in §5.2.

5.1.4. Planetary Interiors

Tectonic processes are clearly of considerable importance on all the terrestrial planets since they are closely associated with degassing of atmospheric constituents, changing topographic features, reprocessing rocks, and hence possibly modification of the chemical state of atmospheric gases, the production of atmospheric aerosols and large-scale plate movement activities which may in turn cause climatic signals via surface albedo changes. The method and timing of core formation and the evolution of a planetary magnetic field are the subject of extensive debate in the literature e.g. Pollack and Yung (1980). Walker (1978c) has suggested that tectonic processes themselves evolve as the planetary interiors cool. Figures 2.3 and 2.4 depict the development of the crust and lithosphere and the evolution of tectonic activity with time. It is possible that atmospheric chemical equilibrium with surface rocks cannot be maintained indefinitely after the periods of terminal vulcanism before final planetary quiescence is achieved. This suggestion is clearly of importance for planetary configurations in which major volatiles are not able to form surface deposits (e.g. the case of Venus). Very much more data on the cytherean topographic features are required before any detailed planetary discussion can be undertaken. For the Earth, McCrea (1981) has established the following characteristic times for tectonic activity—large-scale continental movements $\sim 5 \times 10^7$ years; successive openings and closings of ocean basins $\sim 5 \times 10^8$ years (between 5 and 7 cycles in all during the past 3.8×10^9 years) and major changes in plate motions with periods of several 10^7 years.

It would be interesting to try to establish whether these time periods are themselves functionally dependent upon the state of planetary tectonic evolution as described above. Such a study could be undertaken if Viking imagery for Mars can be interpreted in such a way that characteristic times similar to those listed above for the Earth can be estimated for Mars.

5.1.5. Other Geo- and Astrophysical Perturbing Effects

There are a number of cyclic and quasi-cyclic phenomena which may be regarded as external to climatic systems and

168

separate from the topics discussed above but which would still influence the Earth's environmental regime. Galactic mechanisms could cause both solar and terrestrial changes with periods of the order of 10^8 years (Clark *et al* 1977, McCrea 1981). The precise mechanism by which the passage through dust and gas clouds modifies the climate is, as yet, not fully understood. There is a suggestive correlation between very long-term glacial epochs and this galactic rotation time signature. However Cogley (1981) has re-evaluated the data and finds little evidence for cyclic or quasi-cyclic events. Figure 5.1 suggests that there is, at best, an absence of signature.

Other astronomical effects such as supernovae and galactic explosions may, it is suggested (Wolfendale 1978), modify the Earth's climate either directly (Wolfendale estimates the magnitude of the γ-ray flash at the top of the Earth's atmosphere as approximately $10^6 \, \mathrm{J\,m^{-2}}$) or by perturbation of atmospheric constituents; for instance the stratospheric ozone layer on the Earth may be considerably modified by such high-energy events. Seyfert and Sirkin (1979) suggest that

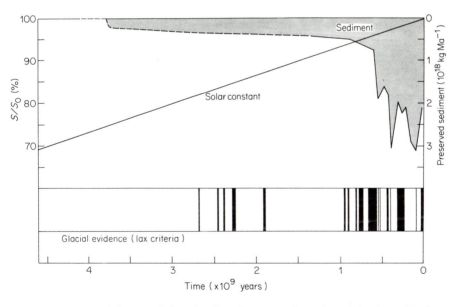

Figure 5.1. Glacial record for the Earth reconstituted mainly from Frakes (1979). The suggestion of quasi-cyclic occurrences of glaciations does not seem to be supported by these data (after Cogley 1981).

meteoritic impacts occur in epochs which have a quasi-cyclic time signature of around 2×10^7 years.

McCrea (1981) has considered in detail the features and periodic changes in the terrestrial magnetic field. The pattern of magnetic field reversals is very complex, but he summarises the published results as (see also McElhinny 1979):

'(*a*)   back to about 4.5 My there has been an average time of roughly

$$2 \times 10^5 \text{ y between reversals}$$

and this may be the significant average also over a very much longer time (McElhinny 1973),

(*b*)   back to about 550 My there has been a characteristic time of about

$$5 \times 10^7 \text{ y between quite abrupt changes}$$

in the frequency of reversals, or changes from one predominant polarity to the opposite (McElhinny and Burek 1971).'

There are also a number of suggestive correlations between geomagnetic events and a large number of the many internal and external phenomena that are listed in this section. However, the careful and detailed review of McCrea (1981) from which many of the data or inferences have been drawn seems to suggest that there may be as yet unconsidered feedback processes operating between the internal and external perturbing factors described.

The direct impact of the weather on the atmospheric environment via lightning discharges and the solution of trace gases through precipitation have already been discussed (chapter 4), and the fundamental effect of extensive liquid water on both the ambient levels of CO_2 (§3.4) and, for instance, ammonia have been considered in detail (chapter 4). There are, however, secondary effects which provide more links between the physical and (bio)chemical characteristics. A very interesting negative feedback effect has been suggested by Walker *et al* (1981) which links the rate of removal of CO_2 through weathering to the ambient surface temperature—enhanced temperatures leading to more rapid removal of CO_2 and stability being retained through a reduced remov-

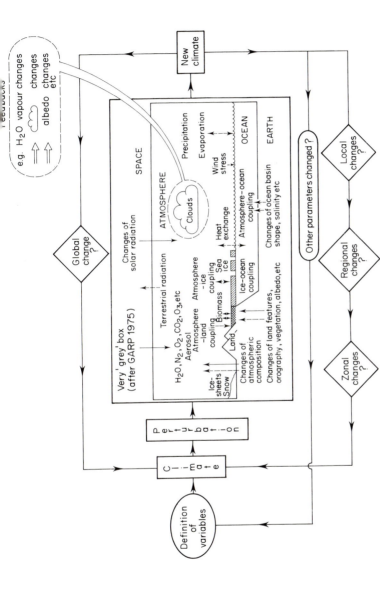

Figure 5.2. The complex interactions amongst the many elements of the climate system on Earth. Feedback processes can magnify, decrease or modify a perturbation so that change may occur either on a global scale or via local and regional reinforcements.

operating on the planet. For instance orography and continental area and position will be functions of tectonic activity. It is also necessary to consider the many timescales of interaction that the GARP (1975) definition implies, for instance the atmosphere responds most rapidly to external changes with a 'weather' relaxation time of approximately a month on the Earth but closer to a day on Mars. When the interactions between e.g. land surface processes (tectonic activity) and states of volatiles e.g. oceanic uptake on the Earth (say by solution of CO_2) or condensation of polar ice caps on Mars are considered, timescales of change may be larger than those of all the other components in the system i.e. evolutionary. In the same way the upper layers of the ocean are known to interact with the atmosphere on timescales of between months to years whilst temperature changes in the deep oceans, ocean basin morphology and sea level fluctuations occur over much longer time periods. The 'levels' of activity within the cryosphere were excellently reviewed by McCrea (1981).

The interactions between the wide range of processes to be considered are of fundamental importance but are not yet well understood. There is, for instance, now good reason to believe that many of the interactions betwen the five components of the *climate system* are non-linear. For example, bottom water temperatures in the Late Cretaceous (approximately 0.07×10^9 years BP) were as much as 15 K warmer than at present. Since cold polar regions imply cold bottom water formation, it is likely that the poles were warm relative to today. Temperature estimates range between 278 K and 292 K. This agrees with other geological evidence that the Earth was relatively ice-free at that time (Hays 1977). Assuming that the atmospheric composition 10^8 years ago was not very different from today, and barring any unsuspected changes in solar output, this large temperature enhancement must be explained by other—probably internal—climatic mechanisms. Qualitatively it has been suggested that the warmer temperatures resulted from the large heat capacity of the submerged continents. Credible, quantitative explanations of this massive, but relatively recent, climatic change are suggested in Barron *et al* (1981) but these authors

174

caution that use of present-day climatic models for palaeo-climatic simulations must be undertaken with care. Their computations include surface albedo changes and a change in the cloud cover which, they postulate, would be a result of the decreased pole-to-equator temperature gradient (see §2.1.3). Sagan *et al* (1973) have suggested that planetary climatic change could be 'triggered' by comparatively small fluctuations in solar luminosity on Mars but it seems that the climatic environment (at least as measured in the simplest way by average global surface temperature and the abundance of surface liquid water) on Earth has remained surprisingly invariant throughout its evolutionary history spanning almost 4.5×10^9 years. Models of planetary climate are unable to incorporate the complex interrelationships shown in figure 5.2. A simple model of interactions within the Earth's climate is reviewed below together with a brief discussion of some of the more important feedback processes.

Early condensation of water on the Earth constrains the evolutionary state of the atmosphere such that the total mass is relatively less variable (§3.4 and the appendix) than either Mars (tenuous and fluctuating) or Venus (rapidly increased and then (?) static). The Earth's surface temperature must be a sensitive function of the net radiation fluxes and the vertical transfer of energy away from the surface (§2.1). It is proposed that the climatology of a hydrospherically dominated planet such as the Earth must be studied somewhat differently from either the tenuous or 'runaway' atmospheric evolutionary histories (Rasool and De Bergh 1970). Here an investigation is made of the sensitivity of the climatic response to forcing effects over evolutionary timescales rather than to evolution *per se*.

Climatic sensitivity parameters are discussed by e.g. Schneider and Mass (1975), Cess *et al* (1982) but these are probably too sophisticated for this discussion which must be constrained to the globally averaged case.

5.2.1. *Climate and the Planetary Albedo*

The analysis presented here is a development of the Budyko–Sellers (Budyko 1969, Sellers 1969) climate model but it

175

differs from other developments of this model type (North 1975) in that a long-term planetary application is desired rather than a zonal climatic predictive mode. The investigation focuses on the absorbed solar radiation, *I*, and the emitted infrared radiation, *F*. The upper graph of figure 5.3

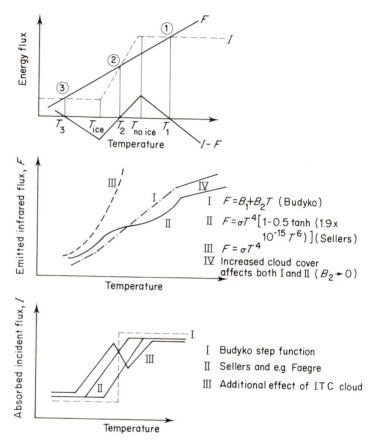

Figure 5.3. Three graphs illustrate the use of a Budyko–Sellers energy balance climate model to predict possible global climatic states. Mean global temperature is plotted against energy flux. The *upper graph* shows three climate states at the intersections of the absorbed flux curve, *I*, and the emitted flux curve, *F*. The stability of these three regimes, say to a change in solar flux, is given by the slope of the *I–F* curve (see text for fuller explanation). The *centre graph* illustrates four methods of parametrising the emitted flux, *F*. The *lower graph* illustrates three possible curves for the absorbed flux, *I*, which varies as a function of the parametrised albedo (after Henderson-Sellers 1982).

illustrates the way in which the intersections between a curve for I, here taken as a function of the simple albedo parametrisation:

$$A = \begin{cases} A_{ice} & \text{for} \quad T < T_{ice} \\ A_{no\ ice} & \text{for} \quad T > T_{no\ ice} \\ \text{linear function} & \text{for} \quad T_{ice} \leqslant T \leqslant T_{no\ ice} \end{cases} \qquad (5.1)$$

and F, define global climate states in terms of the predicted surface temperature, T. The formulation used here for F follows Budyko (1969).

$$F = B_1 + B_2 T \qquad (5.2)$$

where B_1 and B_2 are empirical constants—a form in part supported by satellite observations (Warren and Schneider 1979). Three global climates are determined by the intersection of these curves. Climate 1 is considered to be representative of the present-day situation with climates 2 and 3 being designated glacial and ice-covered with values of $T_1 = 288$ K; $T_2 = 267$ K and $T_3 = 175$ K, respectively (Crafoord and Kallen 1978). Ghil (1976) demonstrates that the stability of the predicted climate is dependent upon the gradient of the $(I–F)$ curve, a positive gradient implying an unstable climate. The effect of a decrease in solar luminosity is easily seen. The I curve would be lowered but the F curve would remain. The two upper climatic states 'approach' one another thus making transition from present-day to glacial conditions easier. Eventually states 1 and 2 vanish and the Earth moves rapidly to climate 3.

Alternatively Sellers (1969) has constructed an empirical relationship from data on zonal temperature and fluxes and suggests

$$F = \sigma T^4 [1 - 0.5 \tanh (1.9 \times 10^{-15}\ T^6)]. \qquad (5.3)$$

Both these equations (equations (5.2) and (5.3)) are depicted in the centre graph of figure 5.3 together with the curve for the limiting case of a planet without an atmosphere:

$$F = \sigma T^4. \qquad (5.4)$$

It should be noted that although T is representative of the

177

average surface temperature, it has not been replaced by T_s in these equations for two reasons. Firstly, much of the early observational work and empirical curve fitting used °C rather than K and hence values of constants must be considered carefully. Secondly, the observations referred to above and upon which equations (5.2) and (5.3) are based are made for zonally averaged surface temperatures rather than individual whole planet configurations. The effects of clouds on the relationship between F and T is very important (the emitted radiation from clouds is very much less than that from the surface or clear sky), but difficult to quantify. A possible effect of clouds upon the F curve parameters has been included in figure 5.3 (for further discussion of cloud–climate feedback see §2.3 and below).

The lowest graph in figure 5.3 illustrates various curves of absorbed solar radiation, I, as a function of temperature, T. Equation (5.1) has been reworked (changing gradients and upper and lower albedo values) by a number of authors (see e.g. Crafoord and Kallen 1978, Schneider and Thompson 1980). Satellite data now available (e.g. Winston *et al* 1979, Ellis, 1978, Campbell and Vonder Haar 1980) indicate that in certain temperature regions, especially in the highly baroclinic zone near the snow/ice boundary and in the Inter Tropical Convergence, clouds dominate the planetary albedo. Thus the model can be improved by the inclusion of these empirical results and the redrawn albedo curve (figure 5.4) permits recalculation of the temperature climates as described below. However, it must be noted that climatic predictions made from even as simple a model as the Budyko–Sellers-type radiation balance is hazardous (Schneider and Thompson 1980). The emitted infrared flux is dependent upon the optical depth and hence gaseous and particulate amounts. The absorbed flux is a function of clear-sky scattering and the surface albedo, but these sources of error are insignificant compared with the potential variability in parametrisation of the cloud–climate relationship (see §2.3).

The pair of flux curves in figure 5.4 are for a planet with partial cloud cover resulting from a well mixed and convecting troposphere (e.g. the Earth). The fascinating features now are the number and stability of the predicted climates.

178

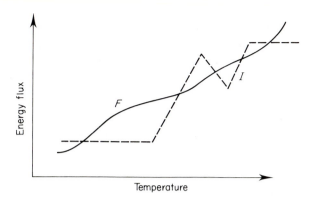

Figure 5.4. Absorbed flux, *I*, and emitted flux, *F*, curves representative of a planet possessing an atmosphere and clouds. The cloud cover modifies both the albedo and the emitted flux. The result is to bring the predicted climate states 'closer', cf figure 5.3, and to reduce the temperature change expected as a result of external forcing (after Henderson-Sellers 1982).

The much closer proximity of the absorbed and emitted curves suggests that the standard three climates predicted by the Budyko–Sellers-type models could easily be replaced by a regime of five, or possibly more, climates. The required additional intersection of the curves is within the error bars on both the system albedos used and the method of incorporating cloud effects into the *F* and *I* curves. Furthermore, if a slight perturbation were to occur in the global cloud pattern (e.g. Roads 1978, Wetherald and Manabe 1980) the predicted temperatures and/or the stability of the 'present-day' and 'glacial' climates would be changed. Perhaps even more important is the fact that all climates except the ice-covered Earth are close together (possibly implying ease of transition) and that even large changes in incident solar radiation will not remove these climatic states. The important point to be underlined is the result that clouds and atmospheric water vapour have a moderating influence upon both the 'distance' between predicted regimes and upon the magnitude of perturbations due to external changes.

These conclusions, whilst being highly tentative, are at least in agreement with the working hypothesis that over long

179

time periods the Earth's climate could be stable to most external variations.

The direction of the cloud–climate feedback is not known (§2.3). Indeed, its net effect is questioned (e.g. by Cess 1976, Cess *et al* 1982) who believe that the effects of perturbed cloud amount upon the albedo and the emitted infrared radiation are mutually compensatory. Certainly the data are, as yet, not adequate to permit conclusive statements about either the sign or the magnitude of feedbacks. Furthermore, it is not only the cloud amount but the cloud type, height, and possibly also the total optical thickness that could control the climatic response (see e.g. Wang *et al* 1981).

Furthermore, Henderson-Sellers (1978) suggests that cloud configuration is a weak function of continent–ocean configuration. Over geological time periods it appears to be important to include cloud feedback effects in climate models. The preceding discussion of I and F flux curves underlines the need to include the infrared as well as the shortwave radiation changes. On a local scale these changes need not be compensatory. For example, a build-up of low-level stratus cloud over high-latitude ocean areas substantially changes the local albedo, but the emitted flux is unlikely to be greatly modified. (Such a change has been proposed as a possible system response which would dampen if not completely remove the positive ice-albedo feedback anticipated in response to anthropogenically increased CO_2 levels.) However, Wetherald and Manabe (1980) suggest that on a global scale cloud changes can become mutually compensatory. It is not clear that results such as these can be extrapolated to evolutionary timescales.

Since cloud–climate responses are so ill understood in the current situation, and since observational data are still distressingly sparse, methods of parametrising clouds for simpler climate models often result in a wide range of responses. It is hardly surprising that the few attempts that have been made to consider the effect of clouds upon long-term climate trends should produce diverse and even opposing results. Hart's (1978) model is extremely sensitive to cloud–climate variations. He assumes that the percentage cloud cover is proportional to the total atmospheric water vapour (which is

180

linked to T_s by the Clausius–Clapeyron equation). Thus his calculated albedos are a very strong function of T_s, whilst he assumes that the dependence of F upon T_s is similar to that for the present day. This provides a strong negative feedback through cloud amount on T_s (Owen *et al* 1979). Alternatively, Henderson-Sellers and Meadows (1979a) suggest that the cloud amount is unlikely to change by a significant amount over geological time periods and thus infer a weak cloud–T_s relationship. Finally, some results from general circulation (GCM) simulations of the current atmosphere (Roads 1978) have indicated a slight tendency to decrease the overall cloud cover in response to increased surface temperatures.

The pair of curves in figure 5.4 is a first, highly empirical, attempt to represent a parametrisation of the cloud–climate relationship in a way that can be used for climate modelling over geological or evolutionary timescales. The cloud–climate problem is potentially more serious for long-term climate modelling than for short-term impact and sensitivity studies, since for the latter the tuned parametrisations in GCM should be adequate. It is not known (i) whether the percentage cloud cover is likely to change, (ii) whether the cloud type and height will vary and (iii) whether all, some, or none of these changes will be exactly compensated for by opposing modifications in the absorbed solar and emitted infrared fluxes. Clouds could be a cause of long-term climate change or could be purely incidental to it. Figure 5.4 suggests that clouds are actually an important moderating influence on the climatic regime.

The two topics discussed in this section (i.e. radiation fluxes and clouds) are clearly not independent of one another. On the contrary, the net global radiation balance can be very strongly modified by the nature, extent and vertical position of clouds. If, as the data described in chapter 4 seem to suggest, there has been a tendency towards global climatic stability on the Earth, then it seems probable that clouds could play an important role. Equation (2.4) can be rewritten in terms of the tropospheric lapse rate (§§2.2 and 2.3) following Hansen *et al* (1981) as:

$$T_s = T_e + \Gamma H \qquad (5.5)$$

where Γ is the mean environmental lapse rate and H (km) the globally averaged height of the radiating layer. If the tropospheric lapse rate is a fundamental and fairly stable feature of planetary systems as suggested by Hansen *et al* (1981), then the position of the mean emitting layer (or the main cloud deck height) must be expected to change in response to climatic fluctuations (compared with the discussion of figure 4.3).

The model presented here may be applied to any planetary system if the absorbed and emitted fluxes can be parametrised. A similar but much more detailed investigation specific to Mars has recently been made by Hoffert *et al* (1981). Their model is particularly useful for shorter timescale changes as it includes explicit calculation of the effect of meridional transport of energy by the atmosphere. They consider two cases: (i) the present martian configuration ($p_s \simeq 7$ mbar) in which the latitudinal transport is approximately zero and (ii) a warmer environmental regime for Mars in which, with a surface pressure of approximately 1000 mbar, water vapour is liberated to the atmosphere and a temperature distribution which will permit liquid water over 95% of the planetary surface is computed. Figure 5.5 illustrates these two temperature regimes as a function of latitude. These surface temperature calculations depend upon the changes anticipated in the infrared flux to space under different atmospheric conditions. Hoffert *et al* (1981) have calculated these fluxes as shown in figure 5.6. These curves compare with those in figure 5.3 (after Henderson-Sellers 1982). Hoffert *et al* (1981) have chosen to neglect the effects of cloud cover in these curves of emitted infrared radiation but the agreement between their calculated values as a function of the mean planetary temperature, T_s, (for $T_s \lesssim 280$ K) and the curves in the centre graph of figure 5.3 is good. Their approach is self-consistent since they also omit the effects of clouds upon the planetary albedo. Hoffert *et al*'s (1981) solution of a $2D$ energy balance problem presented in figure 5.5 demonstrates the importance of including the effects of meridional energy transport in any climate (as opposed to evolution) model. It is especially interesting to note the much more uniform latitudinal temperature profile which they calculate for their 'hot Mars'

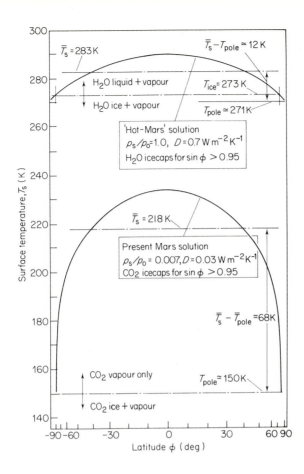

Figure 5.5. Two possible annual mean climatic states on Mars with the same distribution of insolation but different surface pressures of carbon dioxide. The 'hot Mars' solution has water icecaps with an albedo of 0.62; the 'present Mars' solution has CO_2 (dry) icecaps with an albedo of 0.64. Both models assume a bare ground albedo of 0.21 and neglect the possible influence of clouds (after Hoffert *et al* 1981).

example (surface pressure, $p_s \simeq 1000$ mbar CO_2). The importance of meridional energy flux increases very rapidly as the atmospheric pressure is increased. The same mechanism has been examined in much greater detail for the Earth's climate by Barron *et al* (1981) but in this case considerable emphasis was placed upon the importance of cloud cover.

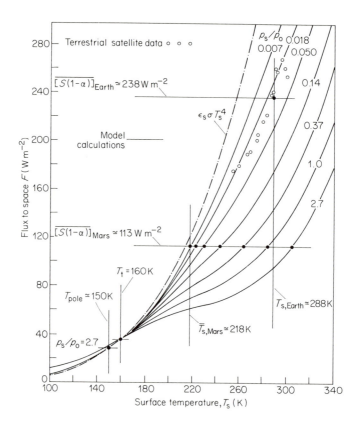

Figure 5.6. Infrared flux, F, to space for a CO_2 atmosphere (includes water vapour feedback) plotted against surface temperature computed with a numerical model for long-wave radiation (see Hoffert *et al* 1981). These curves for F are directly comparable with those in the middle graph of figure 5.3. The global mean surface temperature is defined by the intersection of the $F(T_s)$ curve with the mean solar flux absorbed by the planet, as in the terrestrial example where observational data on $F(T_s)$ are available from satellite radiation and surface temperature measurements (see text). For Mars, a range of surface temperatures is possible which balance the absorbed solar flux, depending on the surface pressure of CO_2 (after Hoffert *et al* 1981).

5.2.2. *Degassing and Planetary Climatic Change*

The imaging missions to Mars of the Mariner 9 and Viking orbiter spacecraft have revealed surface features which are almost certainly due to surface liquid water. There are three types of these channels: (1) runoff channels which are dendritic networks of relatively small channels/valleys and generally appear in the old, densely cratered terrain; (2) outflow channels which resemble large-scale tributaries and (3) fretted channels which are long valleys having flat floors and widening downstream (Spitzer 1980). Dating these features is of critical importance but very difficult (Mutch 1979). It is possible that water floods, which must have been enormous, (peak discharges of 10^7–10^9 $m^3 s^{-1}$ cf an average Amazonian discharge of $10^5 m^3 s^{-1}$) produced the teardrop shaped 'islands' and terraces in the Chryse region. These may have occurred fairly early in Mars' history (say within the first 10^9 years). However, the fretted channels seem to appear throughout the martian geological record. Finally comparatively young features indicate frozen subsurface water remaining; for instance the 'flowing' of ejecta after meteoritic impact has been compared to mud flows on Earth (Carr *et al* 1977). The layered ejecta around Crater Tarsus can even be seen to have flowed around a smaller obstacle. On sections of the Valles Marineris the tributaries to the south side begin in a cirque-like feature which may imply ground water sapping from a frozen subsurface layer of water ice (Farmer and Doms 1979, Spitzer 1980). These features, especially those designated as 'fluvial' seem to imply a very different climate on Mars with a much more substantial atmosphere permitting surface flow of water and possibly even precipitation. The results from the Viking orbiter images indicate that the ages of these fluvial features span most of the lifetime of the planet (Masursky *et al* 1977, Carr 1980) (i.e. at least 3.5×10^9 to 0.5×10^9 years BP). The 'cool-Sun–enhanced-climate' paradox can, it appears, be satisfactorily resolved by allowing an early surface atmospheric pressure on Mars of approximately one bar (Cess *et al* 1980). Indeed the timing and rate of degassing are also important for consideration of the isotopic ratios (especially of nitrogen) found by the Viking entry probe

(McElroy *et al* 1977). Fairly rapid (of the order of 10^4 years), quasi-catastrophic cycles in the climate of Mars may be difficult to reconcile with both the evidence and models for the very early history of the planet and with the current volatile inventory as established by the Viking missions.

Here the effect upon both surface temperature and total volatile inventories of Milankovitch-forced climate change is examined within the framework of the evolutionary history of the atmosphere. The results described are a first attempt to differentiate between degassing modes over the planetary history. The conclusion (that a constant degassing rate is the most satisfactory model of atmospheric evolution) is insensitive to any likely perturbation to the molecular degassing ratio assumed and is consistent with post-Viking results. (Here the ^{40}Ar:CO_2:H_2O ratios are taken to be 1:2000: 40 000 (molecules) which is in accordance with values suggested by e.g. Owen (1976).)

The rate of release of gases to the surface–atmosphere system of any terrestrial planet is extremely difficult to assess. This is because the volatiles released may be lost from the system in a number of different ways (chapter 2). The rare gases (particularly argon) are especially useful as indicators of degassing activity. For instance 99.6% of the argon in the Earth's atmosphere is ^{40}Ar. Once released to the atmosphere this remains as an inert constituent which is extremely unlikely to escape even from a planet as small as Mars (Levine 1978). The nature and timing of the release of volatiles from the surface carbonaceous-type material may not be uniform for the three planets. A very early impact/degassing history seems to be satisfactory for the Earth (§2.1) and indeed an early and rapid evolution of the martian atmosphere has been urged by Fanale (1971) and more recently by an increasing number of researchers e.g. Cess *et al* (1980). However, it also seems reasonable to anticipate that degassing is associated with the level of tectonic activity (e.g. figures 2.3 and 2.4). Mariner 9 data for Mars suggested that tectonic processes were still in a growth stage and more recently Viking imagery has been used to establish that some martian lava flows are geologically young—less than 0.2×10^9 years (Masursky *et al* 1977). Comparison of crater densities suggests that there

186

have existed centres of volcanic activity throughout the life-time of the planet. Thus, despite the fact that there are interesting arguments in favour of an early, rapid degassing, it seems that an evolutionary history in which considerable contributions to the atmospheric mass have been made throughout the lifetime of Mars does not conflict with current geological evidence.

The degassing rate of primary volatiles can be expressed as

$$D(t) \simeq D(t_0)(1 - e^{-\lambda t}) \tag{5.6}$$

where $D(t_0)$ is the total degassed volatiles and $D(t)$ is the degassed mass at time t. Rubey (1951) suggested that the release of volatiles on Earth has been approximately uniform throughout time i.e.

$$D(t) = D(t_0)t/T \tag{5.7}$$

where t_0 is the total age of the planet. Li (1972) considered a series of degassing models for the Earth ranging from this linear mode of Rubey through a family of the asymptotic equations (λ ranged from 1×10^{-9} years^{-1} to infinity) in terms of the geochemical mass balance amongst the lithosphere, hydrosphere and atmosphere. He concluded from Garrels and Mackenzie (1969) that sedimentary mass (as a function of time) does not indicate directly the degassing mode and rate. Following Birch (1965) he selected an asymptotic degassing model.

It is possible that the degassing rate has varied from planet to planet in the solar system. It was for this reason that three degassing sequences were considered in an earlier evolutionary history of Mars (Henderson-Sellers and Meadows 1976). These were: I, an early, almost total, degassing; II, a linear degassing with time over the lifetime of the planet; and III, a late degassing over the last 2×10^9 years of the planet's history. Figure 5.7 shows the average surface temperatures calculated for these three cases. In all three the mean surface temperature of the planet remains below the freezing point of water. However, Mars experiences large temperature fluctuations on two independent timescales both of which are too small to be resolved in an evolutionary history spanning geological periods such as that shown in figure 5.7 (see also

187

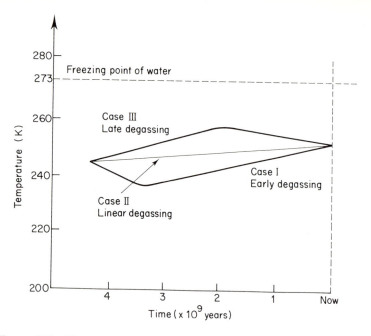

Figure 5.7. The evolution of the surface temperature of Mars for the three degassing modes: case I, early degassing; case II, linear degassing; case III, late degassing. (After Henderson-Sellers and Meadows 1976, see also figure 3.11.) The most likely degassing history cannot be determined directly from the mean planetary temperature curves since climatic excursions are so large. The remaining volatiles (e.g. H_2O as in table 5.2) may permit the selection of the 'best' degassing sequence.

§3.4). These temperature fluctuations are a result of the large changes in insolation which occur on both a diurnal timescale and as a result of the Milankovitch perturbations in the planet–Sun orbital configuration. It is evident from the analysis of the data derived from the Viking mission that case III (late degassing) is unsatisfactory because it cannot explain the earliest 'fluvial' features which date back at least 3.5×10^9 years (Masursky *et al* 1977, Spitzer 1980). It is possible that case I (early degassing) is also unsatisfactory since it cannot easily explain the current surface–atmosphere volatile mass suggested from the Viking data. Table 5.1 lists the estimated mass of water currently in the martian surface–atmosphere system (Sagan *et al* 1973, Soffen 1977). This volatile estimate includes a subsurface water ice component. Even with the addition of this component the total mass of H_2O is clearly in disagreement with the predicted (directly from the observed mass of ^{40}Ar) amount of H_2O degassed which is of the order of 4.5×10^4 kg m^{-2}. If the theoretical calculations are correct there has been a considerable loss of water vapour from the atmosphere. The loss mechanism must fulfil three important criteria: (i) remove the majority of water degassed; (ii) permit extensive climatic variation consistent with 'fluvial' and 'ice age' conditions throughout most of the history of Mars and (iii) retain a small (but non-negligible) mass of water ($\sim$6–11 kg m^{-2}) to the present day.

The long-term evolution of the surface–atmosphere system of any terrestrial planet depends upon solar insolation and the complex feedback mechanisms between the available volatiles and the ambient surface temperature. Superimposed upon this long-term evolutionary trend some of the planets suffer shorter-period cyclic variations. An investigation of the evolutionary impact of such climatic cycles for the planet Mars is presented below.

Many of the planets seem to have surface conditions in the region of the triple point of their major atmospheric constituent (see chapter 6) but only Mars (and possibly Titan) are, at the same time, small enough to have suffered significant gaseous escape. It is for this reason that short-period climatic change must be considered as an integral part of the evolutionary history of these planets.

The astronomical theory of climatic change (now generally called the Milankovitch theory) has been widely discussed. The elegant mathematical calculations of Berger (1977) combined with the data of Hays *et al* (1976) seem to have

Table 5.1. Approximate H_2O inventory for the present-day martian surface–atmosphere system.

H_2O (kg m^{-2})	Part of system
0.009	Atmosphere
0.9	Polar caps
5–10	Sub-surface ice (estimated)
6–11	Total

substantiated the claim that insolation changes are associated with climatic perturbations on the Earth. The situation on Mars is considerably more extreme because the atmosphere (at least at the present time) is more tenuous and the variation in orbital parameters more complex (Ward 1973). For instance, the obliquity of Mars oscillates over two time scales: 1.2×10^5 years and 1.2×10^6 years; leading to a range of obliquity values from 14.9 to 35.5 degrees. It has been suggested that these periodic changes in insolation may be associated with the layered deposits in polar regions (Murray *et al* 1973) and indeed that climatic cycling may result directly from this obliquity driving (Sagan *et al* 1973). Here the analysis follows Sagan *et al* (1973) in considering the changes resulting from insolation variations in polar regions. The flux of solar energy absorbed at the pole is given by

$$S = S_0 a^{-2}(1-e^2)^{-1/2}(\sin \delta/\pi)(1-A) \qquad (5.8)$$

where S_0 is the solar constant at 1 astronomical unit (AU), a is the semi-major axis of the martian orbit, e the orbital eccentricity, δ the obliquity of Mars and A the albedo of the polar cap.

Once the polar insolation attains a value great enough to allow sublimation of all the volatiles bound into the polar cap, it seems likely that a climatic regime considerably different from that observed today will ensue (e.g. Hoffert *et al* 1981). Recent improvements in the calculation of infrared absorption properties of carbon dioxide dominated atmospheres (Owen *et al* 1979, Cess *et al* 1980) and the conclusive evidence for the occurrence of planetary dust storms with the consequent admixture of dust into the polar caps suggest that this threshold value should be less than previously assumed.

It is possible that the mobilisation of large quantities of polar carbon dioxide and polar and subsurface water vapour will occur whenever S achieves a value greater than 27.0 W m^{-2} (approximately 10% higher than the current polar insolation). The considerably enhanced climatic conditions will permit formation of the fluvial features noted from the Viking images but almost certainly also imply higher tropopause temperatures (§2.3) and hence more rapid removal of H_2O by photodissociation of water vapour and the subsequent

190

escape of hydrogen. (The free oxygen resulting from this reaction is probably bound into surface and subsurface layers—the $^{18}O:^{16}O$ ratio in martian CO_2 requires the presence of a surface or subsurface reservoir of oxygen available for interaction with the atmosphere (McElroy et al 1977).) Calculation of the escape rate of hydrogen from this enhanced martian atmosphere has been made for two different exospheric temperatures, namely 300 K and 1200 K. In each case the most probable atomic velocities (2.24 and 4.5 kms^{-1} respectively) compared with the escape velocity for Mars (5.0 kms^{-1}) are such that practically all ($\sim90\%$) available atmospheric water vapour would be lost in a comparatively short time (i.e. less than the time period of polar insolation change associated with obliquity and eccentricity variation). This rapid dissociation of water vapour and escape of hydrogen has been invoked to explain the highly oxidised nature of the martian surface (Horowitz et al 1977). However, this computation suggests that the nature of this loss mechanism imposes an important constraint on the degassing rate for the planet.

In the computations only the loss of water vapour is considered, i.e. carbon dioxide partial pressures vary as a function of the degassing rate only. Using the variations in the orbital parameters of Mars described by Ward (1973), it is possible to monitor the epochs in which polar insolation would pass the threshold value. During each of these epochs the atmospheric evolutionary model has been modified to account for the loss of water vapour.

The variety of surface features observed may indicate a quasi-secular change in the nature of the water-dominated climatic epochs (Masursky et al 1977, Spitzer 1980). The rapid loss of water from the enhanced climatic regime may be responsible for a change in nature of the short-term eras and hence changed erosion features. The exponential degassing curves of Li (1972) (equation (5.6)) are found to be inadequate since either zero or a negligible proportion of the degassed water is retained in the system. Both Rubey (1951) and Li (1972) have suggested the possibility of approximately linear degassing through time (equation (5.7)) and since this appears to be in better agreement with evidence for tectonic

Table 5.2. Calculated amount of water remaining within the martian surface–atmosphere system from 3.5×10^9 years BP to the present day. Time steps are in 10^8 years and calculations have been made for three different atmospheric loss rates, for a linear degassing mode and three atmospheric loss rates, namely 10%, 91% and 99.1%. (NB CO_2 is assumed to be linearly degassed but is not removed.)

Time ($\times 10^9$ years) after planetary formation	Water (kg m^{-2}) % loss rate within each enhanced climatic epoch		
	10%	91%	99.1%
3.5	3.91	2.91	2.90
3.6	7.79	6.81	6.80
3.7	1.20	0.01	negligible
3.8	3.67	2.71	2.70
3.9	7.67	6.71	6.70
4.0	1.16	0.01	negligible
4.1	3.56	2.61	2.60
4.2	7.55	6.61	6.60
4.3	1.12	0.01	negligible
4.4	3.44	2.51	2.50
4.5	7.34	6.41	6.40

activity, model runs for a linear degassing mode have been considered. The results given in table 5.2 are interesting because they satisfy the necessary conditions of retention of H_2O within the system and also have present-day values of the same order as those suggested in table 5.1. It is important to note here that the conclusion is insensitive to the precise loss rate from the atmosphere. This varying nature of surface fluvial features and their recurrence throughout almost the whole of the history of Mars can only be successfully simulated by a linear atmospheric degassing model.

Recently an alternative explanation has been proposed for the outflow channel features. Lucchita *et al* (1981) have compared these features with similar features on the Earth which are the result of ice stream or glacier flow. The problem of scale (cf the need to invoke floods of considerable magnitude) is alleviated if this mechanism is acceptable since the features found on both planets are of similar dimensions. Also it is interesting to note that this mechanism of ice stream/glacier flow could have continued throughout the planetary history (Wallace and Sagan 1979) whilst, as described above, even under the extreme hypothesis of linear degassing throughout the 4.5×10^9 year planetary evolution the nature of any possible 'fluvial' epochs is considerably modified by the loss of water vapour. The wealth of imagery produced by the Viking mission may permit dating of some of these features which may, in turn, establish the superiority of one of these histories. However, the greater eccentricity of the martian orbit (Ward 1973) suggests that climatic excursions such as those described by Hoffert *et al* (1981) (figure 5.5) will always be a possibility.

5.2.3. *Short-term Chemical Changes*

Recently considerable attention has been paid to the possibility (and predictability) of short-period climatic changes on the Earth and on Mars, but these changes are assumed to result from changes in the physical characteristics of the planet (e.g. albedo, solar radiation). In §2.2.2 it was argued that most chemical (and particularly photochemical) reactions would occur on a timescale too short to be considered important except for their net result for the evolutionary histories described here. However, the planet Titan seems to present an interesting intermediate situation. It must be classed as a terrestrial-type planet since it possesses a substantial atmosphere interacting with the surface. However, its distance from the Sun and interaction with its parent body may result in photochemical and escape processes which operate on time periods similar to, and therefore of fundamental significance for, the evolutionary climate.

Titan is believed to 'recycle' some of its atmosphere (Strobel and Yung 1979) through the hydrogen torus which encircles Saturn close to its orbit. A more interesting weather or climate effect which could have significant consequences for the overall atmospheric composition and structure is that organic molecules are likely to have been 'raining out' of the atmosphere of Titan for much of its history. Smith *et al* (1981) suggest that the surface of the planet may be a methane ocean or frozen surface rich in organic material. It seems reasonable to surmise that such a significant removal

193

mechanism must have affected the chemical composition of the atmosphere but much more detailed study is required before even a qualitative chemical evolution is possible.

Pollack *et al* (1980a) make a brief examination of short-term changes on Titan. They note that Titan's albedo seems to vary by about 10% over time periods of approximately 10 years and that the phase of this global scale planetary change is correlated with the phase of solar activity (Lockwood and Thompson 1979). It is possible that these changes in reflectivity result from dynamical variations in the formation of atmospheric aerosols. Such secular changes in the nature of the atmospheric haze layer would, in turn, modify the amount and nature of solar radiation absorbed and hence the surface temperature.

In this chapter shorter-term climatic changes which may be of importance for the evolution of the atmosphere–planet system have been considered. Little attention has been paid to other short-period fluctuations except for the Earth. This is because the Earth's atmospheric evolution seems to have placed it in a configuration which is remarkably stable. The nature of this stability is examined in chapter 6.

6. Stability of Planetary Environments, Exobiology and the Complexity of Evolutionary Processes

'Life

A common state of matter found at the Earth's surface and throughout its oceans. It is composed of intricate combinations of the common elements hydrogen, carbon, oxygen, nitrogen, sulphur and phosphorus with many other elements in trace quantities. Most forms of life can instantly be recognized without prior experience and are frequently edible. The state of life, however, has so far resisted all attempts at a formal physical definition.'

J E Lovelock 1979

6.1. Climate Models and Evolutionary Timescales

After the upheavals of planetary and atmospheric formation, the climate of a planet may take any one of three paths: (1) homogeneous in time, (2) recovering, and (3) continuing catastrophes. Climate history (1) can be easily dismissed for the Earth as data from the geological record demonstrate that considerable changes have taken place in both the mean and local surface temperatures and in the atmospheric chemistry. Indeed, the evolution of life has almost certainly perturbed the climate (chapter 4 and figure 5.1). It is possible that Titan and Venus may have had climatic histories more like that of path (1) once stability was established and that Mars resembles history (3). Numerous early models of the Earth's climate system (e.g. Budyko 1969, Sellers 1969, Hart 1978) predicted an abundance of climatic catastrophes. It appears that the Earth's 'sister' planet, Venus, may have suffered an irreversible catastrophe.

An extreme 'catastrophe syndrome' for the Earth would, as indicated, imply that environmental discontinuities are acceptable to a biosphere and compatible with the available data. This is contrary to the facts (see chapter 4) although mild climatic 'excursions' may well have occurred. The danger of drawing 'cosmic' conclusions from climate modelling has been excellently described by Schneider and Thompson (1980). Despite the sophistication of current climate models for the Earth there is still considerable uncertainty about basic feedback mechanisms such as those relating to the cloud–climate interaction (Cess 1976, Roads 1978, Wetherald and Manabe 1980) and the appropriateness of convective adjustment schemes and lapse rate choice e.g. Wang et al (1981). It could be argued that the long-term stability of the Earth may be a reflection of negative feedback mechanisms (chapter 5) which are stronger and more effective than has generally been recognised.

In this work planetary atmospheric evolution for all the planets has been reviewed and more detailed data and theories pertaining to the climatic environments of the terrestrial planets considered. From the preceding chapters it is clear that there are both long-term or evolutionary trends and shorter-term stimuli operating in both an external and internal sense within all the climatic systems. It is also evident that the Earth possibly presents a uniquely robust system in which as yet unexplained feedback processes must be compensating for these considerable perturbing influences. Table 4.2 lists the basic characteristics of the Earth. It could be suggested that our climatic system must be viewed in concert with the two other characteristics of our planet which differentiate it from other members of the solar system: the biosphere and the hydrosphere. Lovelock (1975, 1979) believes that global climatic stability is a direct result of the presence of life on the Earth. Alternatively Henderson-Sellers (1981) has proposed that certain of the stabilising features of the planetary system are the direct result of the existence and persistence of a global hydrosphere.

It is just possible that a purely external perturbation to the climatic regime could produce simultaneous and compensatory positive and negative impulses; for instance the passage of the solar system through an interstellar dust cloud might both increase solar luminosity (Clark *et al* 1977) and by enhancing planetary albedo (either through increased atmospheric aerosols or increased cloud cover) reduce the absorbed fraction of incident solar flux (Bethoux 1978). Even in this somewhat unlikely event the final climatic state depends upon the internal responses (e.g. cloud condensation processes). In general it seems reasonable to assume that all control mechanisms will be internal in nature (in the sense used in chapter 5). It is possible to subdivide these major climatic parameters into categories similar to those used in chapter 5 (see figure 5.2): (i) planetary albedo—clouds via the tropospheric lapse rate; (ii) atmospheric—gases and aerosols; (iii) land, ocean and cryosphere surfaces—albedo and biomass; (iv) tectonic processes—plate movement; (v) general circulation processes of both the atmosphere and the oceans. It is possible that the distinction between internal and external is less important than the direction and magnitude of the effects once triggered. Perturbing/restoring feedback effects have been considered throughout but the energy transports within the Earth's system may warrant further discussion.

The Earth's atmosphere is in a permanent, and often turbulent, state of motion. This motion performs the upward and poleward transfer of energy necessary to correct the naturally occurring radiation imbalances (see figure 2.5). This energy transfer is accomplished by the transfer of sensible and of latent heat (i.e. the motion of the atmosphere and the flux of water vapour is of considerable importance) and by ocean currents.

It is only with the comparatively recent advent of global monitoring systems that theories of the likely response of the

197

climate to perturbations in the parameters which force the circulation of energy on the Earth have been able to be evaluated (Dalfes *et al* 1979, Barron *et al* 1981). In epochs at which the pole-to-equator temperature gradient and/or the vertical lapse rate on a zonal or global scale were perturbed the circulation of both the atmosphere and the oceans would respond. Recently Schneider and Thompson (1980) have suggested that the response takes a negative form such that, for instance, an increased pole-to-equator temperature gradient results in enhanced circulation, possibly returning the system to equilibrium (see chapters 4 and 5). Hoffert *et al* (1981) have considered a similar mechanism for Mars. They find that meridional transport of energy is a function of atmospheric mass and results in a considerably decreased equator-to-pole temperature gradient (see figure 5.5).

The complexity and interdependence of the feedback processes operating within the Earth's environment make assessment of the impact of individual events particularly difficult. Ghil (1976), Namias (1978) and Berger (1979) all cite specific examples of non-linear climatic responses operating on very different time periods within the Earth's environmental regime. It is possible that all planetary-atmosphere systems respond in a highly non-linear manner to perturbations. Any attempt at comparison and interpretation of feedback effects is made still more tentative by the fact that (as described in chapter 5) most estimates of the climatic response on the Earth are from *model predictions* rather than observational data. Table 6.1 (from Pollack 1979) is just such a collection of model estimates. It serves, however, to permit first-order comparison of the magnitude and sign of most of the climatic perturbations that have been considered for the Earth. These estimates, of the variations required to cause a 1 K temperature change, are significantly smaller than most similar calculations for other terrestrial planets e.g. Sagan *et al* (1973), re-emphasising the damped nature of the climatic response on the Earth.

Table 6.1 also lists the characteristic time periods of these perturbing effects, hence permitting at least a preliminary comparison between these pseudo-internal feedback factors and the type of external and internal modifications described

Table 6.1. Estimates of the sensitivity of the Earth's climate system to perturbations in various internal and external 'factors' (after Pollack 1979). The variation, ΔV, in each factor, believed to cause a 1 K increase in the mean surface temperature, is listed together with the characteristic timescales of variation. (NB all estimates are derived from $1D$ radiative–convective climate models, see Pollack (1979) for sources.)

Factor	Current value	ΔV	t (years)
Solar luminosity	1360 W m^{-2}	+11.86	10^9; 10^2–10^3 (?)
Oxidation state of atmosphere	Fully oxidised	Partially reducing	10^8–10^9
Continental drift			10^7–10^8
Orbital eccentricity and axial tilt and orientation			10^4–10^5
Volcanic aerosols' optical depth	varies ($\sim$0.005 to 0.01)	-0.12[a]	1–10^6
Tropospheric aerosols' optical depth	0.1	?[b]	[c]10–10^2
CO$_2$ mixing ratio	330 ppm	+125	[c]10–10^2
Ozone column density	0.350 cm atm	+0.55	[c]10–10^2
Stratospheric H$_2$O mixing ratio	3 μg g^{-1}	+6	?
N$_2$O mixing ratio	0.28 ppm	+0.64	[c]10–10^2
CH$_4$ mixing ratio	1.6 ppm	+8	[c]10–10^2
NH$_3$ mixing ratio	0.006 ppm	+0.07	[c]10–10^2
Freon mixing ratio	0.0002 ppm	+0.01	[c]10–10^2
Surface albedo	0.1	-0.01	[d]10–10^5

[a] An increase in volcanic aerosols causes a net cooling.
[b] Even the sign of the effect is uncertain at present.
[c] Inadvertent change due to man.
[d] Either a primary drive, if due to man, or a feedback effect involving a change of polar ice cover.

in chapter 5. Assessment of timescales is almost impossible within this highly interdependent system. Furthermore, the complexity and intimacy of the processes operating in the Earth's climatic environment makes the categorisation of secondary feedback links very difficult and possibly unhelpful. An example of one 'trigger' mechanism is associated with longer-term processes finally leading to a signal causing glacial/interglacial changes. Recent results suggest that the level of carbon dioxide in the Earth's atmosphere was considerably lower (by $\sim$20 ppm) during glacial epochs compared with the present day (335 ppm). This decreased level of CO$_2$ would tend to lower average surface temperatures by decreasing the 'greenhouse effect'. However, the lowered CO$_2$ level may itself be a result of the onset of continental glaciations.

Broecker (1981) believes that when sea levels fell as the ice mass on the northern hemisphere continents increased there may have been a sudden input of phosphorus to the oceans due to leaching from the then dry continental shelves. The increased phosphorus levels would stimulate increased photosynthetic activity in the oceanic biomass and thus lead to lowered levels of CO_2 in the atmosphere. This mechanism provides a link between northern hemisphere glaciations and southern hemisphere temperatures, and also indicates the importance of internal responses within the biosphere and hydrosphere for the enhancement of climatic anomalies. It does not, however, explain the origin of the decreased temperatures that triggered the original continental glaciations.

Furthermore algal productivity can only increase if the C:N:P ratio is close to 106:15:1 (the Redfield ratio; see Redfield 1958) and there is sufficient light. Modern ecological paradigms suggest that N and P levels are unlikely ever to have been high enough to result in the CO_2 algal limitations and hence atmospheric depletion.

The possible uniqueness of the hydrospherically controlled system on the Earth and the lack of planet-wide observational data for other systems makes consideration of these secondary control features impossible for any other planet.

6.2. Life Within and Beyond the Solar System

Over the past decade it has been recognised that interstellar space contains highly complex molecules as well as the simpler atoms and ions which had been anticipated. It has been suggested (Greenberg 1982) that the majority of the oxygen, carbon and nitrogen existing apart from stars is bound up in the form of organic molecules in interstellar dust grains. Thus many of the precursors of life are freely, and even abundantly, available in space. It cannot yet be demonstrated, although it is often stated, that significant amounts of such organic molecules could survive the processes involved in planetary formation (Irvine *et al* 1981, Ponnamperuma 1982). Indeed, on the contrary, it seems unlikely that the traumatic accretion, bombardment, separation and degassing phases

200

which must have taken place on the terrestrial planets could have failed to have destroyed most, if not all, complex pre-biological molecules (Lazcano-Araujo and Oro 1981). Thus the existence of a biosphere seems to require the evolution *in situ* of the precursors of life and finally of life itself (Schwartz 1981).

Contrary to many early hypotheses, geological data for the Earth now seem to suggest that the time periods over which life originated and evolved into a recognisable form and possessing environment-modifying capabilities must be comparatively short—say less than 0.5×10^9 years (and possibly less than 0.2×10^9 years). Hence at best an alien biosphere could have evolved in a relatively short hospitable epoch. It is important to note (see e.g. Schwartz 1972, Rambler and Margulis 1980) that micro-organisms on the Earth exhibit considerable environmental tolerance by three basic types of adaptation to extreme conditions: (i) shielding—the development of a mechanism with which to exclude the harmful factor from the organism; (ii) neutralisation—development within the organism of a method for detoxifying and/or development of rapid reconstruction processes; (iii) toleration—complete adaptation to the harmful factor. It is possible that these adaptive processes may also have assisted the persistence of life elsewhere.

Although the Viking biology experiments produced a considerable flurry of research literature and indeed some tantalising and even suggestive data (Kuhn *et al* 1979), the general consensus is that there is very little evidence for present or past life on Mars. This belief is based upon the absence of organic carbon on the surface. A theoretical model of the martian atmosphere has been used by Kuhn *et al* (1979) to investigate the likelihood of survival of terrestrial organisms in the martian environment. The results of their simulations, which included variations in solar radiation, wind speed and direction and convective, conductive and evaporative cooling in an equatorial surface site, suggest that a primitive community of photosynthetic micro-organisms could survive. They calculate that likely photosynthetic rates could lead to the production of 10^{-3}–10^{-2} moles of O_2 m^{-2} per day.

The Voyager missions to the jovian planets have confirmed

the presence of many organic molecules in the atmospheres of these planets and of their moons. Lovelock (1975) believes that, despite the abundance of these organic compounds, any environment (e.g. the Jupiter atmosphere) which lacks a free-energy gradient is unable to offer an environment suitable for the origin of life itself. In any case it is generally suggested that the evolution of intelligence and of technology are very much enhanced in terrestrial-type situations in which the development of limbs, the necessity to overcome gravity and the development of tool-making facilities and the control of fire are necessary (see Shklovskii and Sagan 1966). Thus it is possible that the existence of *Homo Sapiens* on the Earth may be a direct result of the nature of the rugged surface and the hydrospheric and atmospheric stability.

Margulis and Lovelock (1974) consider in detail the evolution of alien biospheres and particularly the likely gaseous mixture which could be used to identify biospheric systems in operation. Lovelock (1975, 1979) has developed the idea of the importance of inter-reactions between physical and biological systems into a suggested method of detection of alien biospheres. He argues the presence of a mature biosphere is likely to disturb physical parameters in a planetary context to such an extent that the disequilibrium could be detectable from afar. He suggests that strong evidence for a life system is the redox potential of the planetary atmosphere together with an observable level of chemical disequilibrium. For instance the existence in the present Earth's atmosphere of the partial pressure of methane given in table 2.4(a) requires a flux of 10^{12} kg per year together with an equally significant flux of 4×10^{12} kg of oxygen (see figure 4.4) being removed by methane oxidation. Such fluxes which would otherwise require the complete recycling of *all* crustal carbon every 10^8 years clearly demonstrate the existence of a biosphere.

Despite some recent enthusiasm for panspermia, it is generally believed that life on Earth originated *in situ*. This belief immediately places constraints upon the environmental conditions at the time of origin and indeed throughout the evolutionary process (see e.g. Kreifeldt 1971).

The atmospheric characteristics presented in chapter 4 appear to be in conflict with experimental research conducted

202

by many biochemists working on the origin of life. However, the global scale physical and chemical state may not exclude the conditions believed to be necessary for the biochemical building processes. Henderson-Sellers and Schwartz (1980) have described the compatibility of different local and global scale conditions. Locally they have fairly high levels of atmospheric and absorbed NH_3, but the global mixing ratio of NH_3 remains below the level at which climatic effects would be seen (see §4.2.7). This study provides a clue to the resolution of the apparent conflict between astrogeological theories of neutral planetary atmospheres and the laboratory results of biochemists. The climatic feedbacks described in chapters 4 and 5 appear to provide a global situation similar to the present-day Earth. Fairly stable global conditions contrast with highly diverse local environments. It is possible that the diversity of local chemical disequilibria on the surface, in oceans and lakes and even in the atmosphere is a requirement for initiation or nurturing of the evolutionary process of life (Schwartz 1981). Furthermore, weathering and evaporation processes could be of importance in establishing locally anomalous levels of elements and compounds. For instance, peptides have been successfully built from amino acids in experimental environments possessing considerable temperature and moisture fluctuations (Lahav *et al* 1978).

Atmospheric characteristics are clearly important in the provision of hospitable environmental conditions for the genesis and evolution of life; provision of source areas of e.g. abiologically produced NH_3; as a secondary factor controlling gaseous levels through weathering; and direct weather effects, for instance, solution of atmospheric gases as a result of or enhanced by precipitation processes and the direct formation of trace gases by lightning. A further, probably minor, effect is the formation and deposition of trace compounds directly as a result of meteoritic and cometary impact (Henderson-Sellers 1977, Lazcano-Araujo and Oro 1981).

The possible range and chemical nature of life forms has been discussed for some time (Oparin 1938, Shklovskii and Sagan 1966). It appears that the basic 'building block' could be either the carbon or silicon atom though there are excellent reasons for the selection of the former. The presence and fundamental importance of water in all living cells (water composes 80% or more of most cells) may suggest that cellular structure (and hence evolution) will be dependent upon the existence of liquid water within the environment although other elements (e.g. phosphorus) are also believed to be essential (Griffith *et al* 1977). Furthermore the lack of an alternative molecule which possesses as many unique or extreme chemical and physical properties which aid the processes of life itself (e.g. Bayley and Morton 1979) may lead to the belief that water is a basic requirement of life.

This brief and highly tentative perusal of necessary and sufficient conditions for the origin and evolution of life suggests that attention should be focused upon the surfaces of the terrestrial planets which support environmental conditions including surface liquid water.

Schneider and Thompson (1980) have suggested that the 'solar habitable zone may be defined as that region in which a terrestrial planet can retain a significant amount of liquid water at its surface. . .'. The basic planetary characteristics sought should include a mass sufficient to retain a significant degassed or volatilised atmosphere and *initial formation at such a distance from the parent star that rapid surface condensation of water occurs*. It is interesting to note here that another of the extreme characteristics of the Earth is the size of its satellite (table 4.2). Henderson-Sellers and Meadows (1977) have suggested that the effect upon the planetary flux factor, f, of rotation rate changes associated with gravitational interaction with such a large satellite could be an important secondary factor in early values of the surface temperature. This emphasis upon initial formation distance only is contrary to suggestions of e.g. Hart (1978) that the continuously habitable zone (CHZ) is primarily a function of stellar evolution and upheavals in atmospheric chemistry.

6.3. Epilogue—the Uniqueness of Individual Planetary Systems

The evolution of science, like the evolution of life, seems to require a suitable environment and stimulation as well as careful nurturing. It is dangerous, but sometimes useful, to

204

venture forward despite a dearth of data. It is certainly important to state when scant substantive evidence exists. The arguments in chapters 3–6 are at best 'a stab in the dark' probably closer to a 'feeble groping'. However, if these discussions are to further our understanding, there must be an attempt to extrapolate from such hypotheses towards (hopefully) constructive feedback into the other disciplines concerned with the problems of planetary evolution.

I have suggested that the apparently robust nature of the Earth's climate system in the face of atmospheric and surface changes of considerable magnitude may be traced back to negative feedback effects which are often under hydrospheric control (see also Henderson-Sellers 1981). The existence of surface liquid water on Earth and the associated global hydrological cycle seem to be the dominant factors in environmental feedback processes. The inference which must be made from table 4.2 and the discussions in §§3.4 and 6.2 is that the pre-biotic and early biotic Earth had a closer resemblance to the present-day state than previously believed. I have incidentally indicated that the origin and evolution of life (if indeed it occurred on the Earth) probably presented the climate–environment system with perturbations larger than any suffered since (including present-day anthropogenic pollution). Certain of these conclusions warrant further elaboration.

Physical and chemical systems tend towards equilibrium. This has important consequences for the most interesting class of planetary atmospheres, namely those which are both substantial and interact with a surface. The critical feature in these systems, as has been mentioned throughout this text, is the interaction between surface temperature and a phase state change of an atmospheric constituent. Saturation vapour pressures depend exponentially upon temperature. Thus very small changes in temperature which may be a result of small global scale excursions (chapter 5) or local or seasonal variations can result in very large changes in the total mass of the gas in the atmosphere. The data presented here seem to indicate that all the terrestrial planets with atmospheres and Titan possess this feature, although on Venus the surface–atmosphere link is more complex. The dominant atmospheric volatiles are water for the Earth, carbon dioxide for both Venus and Mars and methane for Titan.

There is another similarity between the two planets for which latitudinally resolved data and model results are available, namely that the latitudinal extent of the solid phase of the volatile is approximately the same. Hoffert *et al* (1981) find that their simulation of a 'warm' Mars (§5.2) results in very much greater meridional transport of energy and finally in polar ice caps extending to approximately 72° in close agreement with the analogous situation on the Earth today. Despite these similarities only the Earth seems to have maintained climatological stability over evolutionary history—our planet has been described by Lovelock (1979) as 'a strange and beautiful anomaly in our solar system' (table 4.2).

I sought order in histories of the atmospheres of the planets following the principle of Occam's razor but found it only for our own planet. The Earth seems to be unique in many ways, not least in presenting a picture of stability of the climatic environment. The hydrosphere appears to be fundamental for many of the other unique features. Since all living cells are mostly water they probably evolved in an intimate relationship with water. Carbon dioxide is as important in controlling surface temperature as water vapour but in the absence of a hydrosphere the final equilibrium state is not hospitable. Goody (1975) suggested that 'studies of a single, complex system are often unwittingly constrained by an acceptance of what is observed as "natural" and even "obvious" so that questioning is dulled and scientific enquiry suffers'. I hope that the discussion of the contrasting evolutionary histories of the solar system planetary atmospheres presented here will serve to foster fruitful comparison and to stimulate research.

Appendix. Planetary Temperature Isopleths: Diagrammatic Results From a Numerical Model

Planetary temperature histories can be followed on the graphs showing computed lines of equal temperature. These isopleths of surface temperature (K) are displayed as a function of incident solar flux and atmospheric absorber concentration in figures A1–A5. The value of albedo, A, flux factor, f, and surface infrared emissivity, e, are taken from table 3.3, reproduced below as table A1. The addition of water vapour to the model atmospheres leads to a temperature discontinuity. This has been indicated by a break in both the vertical axis and the isopleths themselves. The three model planets considered bear a significant relationship to the three terrestrial planets possessing atmospheres. The generality of this method of presentation is underlined but comparison can be made between these model evolutionary histories and those for Venus, the Earth and Mars (presented in §3.4). Each planetary evolution is denoted by a numbered symbol (triangle, cross, circle or square) and is described in detail below.

Model 1 △

(i) *Figure A1* (*Upper graph*). The initial conditions for this model are a flux of $2 \times 10^3\,\mathrm{Wm}^{-2}$ (this is a little higher than the value of incident flux at Venus early in its evolution). The planet is assumed to have a flux factor of 2 and an albedo of

Table A1. Value of infrared emissivity, e, for various (likely) combinations of the flux factor, f, and albedo, A, (see also figures A1–A5).

Albedo, A	flux factor, f			
	2.0	2.5	3.0	4.0
0.07	0.90	0.90	—	0.90
0.17	0.93	0.93	0.93	0.93
0.30	—	—	0.95	0.95
0.70	—	—	—	0.95

7% (the surface infrared emissivity is $e = 0.90$). Thus $T_e = T_s \simeq 360$ K ($\triangle$ 1).

(ii) *Figure A5 (Upper graph).* Such a high initial temperature would almost certainly lead to rapid degassing of both carbon dioxide and water vapour. The atmosphere would be rapidly established (so rapidly that the flux has been held constant). The flux factor now has a value of 4 and the albedo is much higher (probably close to 30%) due to the large amount of water vapour available. Thus we find $T_s \simeq 340$ K ($\triangle$ 2).

(iii) *Figure A5 (Lower graph).* Surface temperature has remained high even without flux increase. As the Sun evolves, and the planetary temperature rises, more water vapour will be degassed and evaporated into the atmosphere. The evolution will almost certainly 'runaway', finally producing a situation beyond the range of these graphs, but approximated by the last point: $T_s \simeq 500$ K ($\triangle$ 3).

Model 2 $\bigcirc$

(i) *Figure A1 (Upper graph).* The initial conditions for this model are similar to those for Model 1, except that the flux is lower (taking a value of 1×10^3 Wm^{-2}). This leads to $T_e = T_s \simeq 300$ K ($\bigcirc$ 1).

(ii) *Figure A3 (Upper graph).* Early temperatures above 273 K, together with a slowly increasing flux incident on the planet, leads to considerable degassing together with the build up of an atmosphere of carbon dioxide with small

208

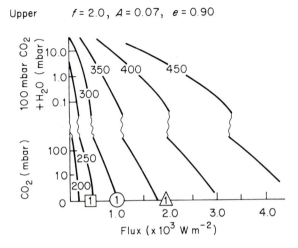

Upper $f = 2.0$, $A = 0.07$, $e = 0.90$

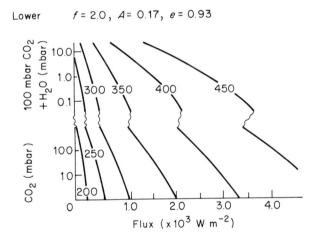

Lower $f = 2.0$, $A = 0.17$, $e = 0.93$

Figure A1. Isopleths of calculated planetary surface temperature as a function of solar radiation (incident at the planet) and atmospheric state (vertical axis). The values of the albedo, A, flux factor, f, and infrared emissivity, e, (from table A1) are indicated. Temperature evolutionary histories of a series of 'model planets' are described in the text.

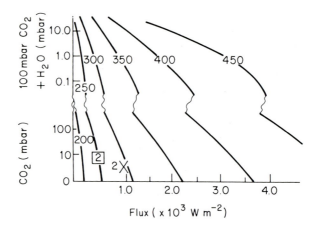

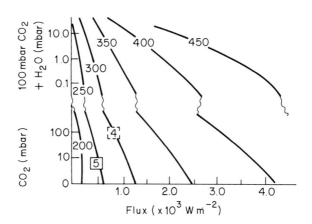

Figure A2. As for figure A1.

amounts of water vapour. The flux factor has a value of 3 (for this intermediate atmosphere). Freezing out of the volatiles around polar regions, or in clouds, leads to an albedo value of approximately 17% and an increased surface infrared emissivity of $e = 0.93$. These conditions produce $T_s \simeq 325$ K (○ 2).

(iii) *Figure A3 (Lower graph)*. Temperatures remain above 300 K (but do not rise rapidly). The atmosphere would probably continue to evolve in a stable fashion under the

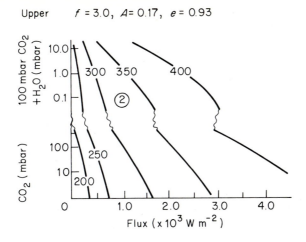

Upper $f = 3.0$, $A = 0.17$, $e = 0.93$

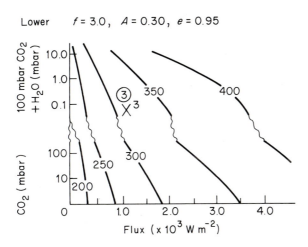

Lower $f = 3.0$, $A = 0.30$, $e = 0.95$

Figure A3. As for figure A1.

influence of increasing solar luminosity. The albedo might increase a little, leading to a new value of $T_s \simeq 320$ K (○ 3). (iv) *Figure A5* (*Upper graph*). Continuing atmospheric build up would produce an increase in f (to a value of 4). Thus we have $T_s \simeq 310$ K (○ 4).

The important and interesting factor in this evolutionary sequence is the way in which increases in incident flux have been adequately compensated for by varying planetary characteristics. The surface temperature has remained very

211

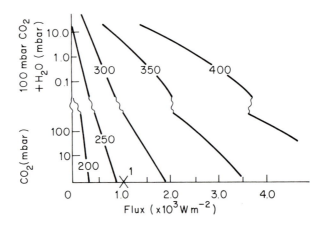

Upper $f = 4.0$, $A = 0.07$, $e = 0.90$

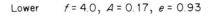

Lower $f = 4.0$, $A = 0.17$, $e = 0.93$

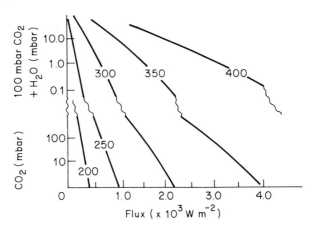

Figure A4. As for figure A1.

stable. However, this evolutionary path is unresponsive to certain features of planetary evolution. For instance, Model 2 evolution has produced stable temperatures in the range 300–325 K together with large amounts of water vapour. These would almost certainly lead to surface condensation of liquid water, and possibly to loss of atmospheric carbon dioxide by solution in water and deposition of carbonates. This has happened during the evolution of the Earth, and the remaining atmospheric CO_2 contributes far less to the 'green-

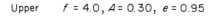

Upper $f = 4.0, A = 0.30, e = 0.95$

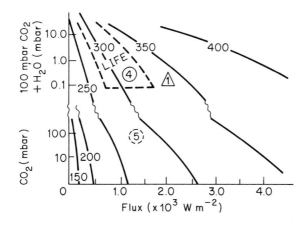

Lower $f = 4.0, A = 0.70, e = 0.95$

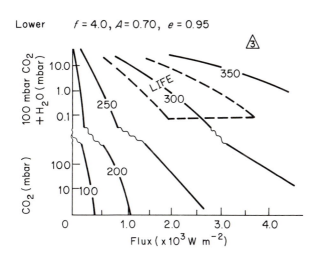

Figure A5. As for figure A1. Additionally areas within the dashed envelope curve show planetary positions for which conditions likely to be hospitable to life may exist.

house' increment than the 100 mbar standard for this model. A further calculation has been made for Model 2 allowing for this loss of CO_2.

(v) *Figure A5 (Upper graph :.).* A slight further increase in incident flux, together with the probable decrease in atmospheric carbon dioxide discussed above, renders the vertical

213

scale on the graph inapplicable. For this reason, the resulting temperature (almost identical to that of the present-day Earth) has been indicated by a dotted circle. It is found that $T_s \simeq 290$ K ($\because$ 5).

Model 3 □

(i) *Figure A1* (*Upper graph*). Initial conditions are similar to those of Models 1 and 2, but the incident solar flux is considerably reduced. It is 0.4×10^3 Wm^{-2}. These conditions lead to a temperature of $T_e = T_s \simeq 245$ K (□ 1).
(ii) *Figure A2* (*Upper graph*). The low value of surface temperature will tend to hamper degassing—probably only of carbon dioxide. A slight increase in atmospheric mass may raise the value of the flux factor to (say) $f = 2.5$. This, together with a small increase in incident flux, results in a very slight temperature change: $T_s \simeq 249$ K (□ 2).
(iii) *Figure A2* (*Lower graph*). Continuing low temperatures combined with larger atmospheric mass may lead to freezing out of CO_2 around the polar regions, and thus to an increased albedo. This almost compensates for the slowly increasing flux, giving $T_s \simeq 250$ K (□ 3).

Two most interesting points are obvious here. Firstly, this evolution (like that of Model 2) seems to be stable. The surface temperature varies little, although the parameters affecting it are changing. This may indicate that a stable surface temperature is the normal planetary condition rather than, as previously generally assumed, a phenomenon requiring complex explanation (see chapter 5). The second point is that this final evolutionary track, which resembles that of Mars, seems to be continuing in its stable mode well below what are termed 'hospitable' conditions. An interesting prediction (attempted by Sagan and Mullen 1972) is the length of time required before a Mars-type planet will evolve to an Earth-like state (assuming, of course, that the necessary volatiles are available). To increase the surface temperature of Model 3 from the last calculated value ($T_s \simeq 250$ K) to a value which would ensure degassing of water vapour and its partial evaporation into the atmosphere (say $T_s \simeq 300$ K) requires a huge increase in input flux. No planetary charac-
214

teristics are likely to change to assist the temperature increase. This calculation has been performed for Model 3.

(iv) *Figure A2* (*Lower graph* :::). To increase the surface temperature to $T_s \simeq 300$ K requires an increase in the incident flux from approximately 0.5×10^3 Wm^{-2} to 10^3 Wm^{-2}. The length of time required to produce this increase for a planet at the same distance as Mars is over 15×10^9 years. This is considerably longer than the time period for which the luminosity curve is applicable, and probably longer than the lifetime of the solar system. However, the long timescale implies that, unless the evolutionary sequence for Model 3 is not applicable to Mars (they could differ if Mars is found to be suffering an intense glaciation at present), the chances of Mars naturally becoming hospitable for advanced forms of life seem to be negligible. (The evolution of life is also discussed in chapters 4 and 6.)

One further parameter variation has to be considered. The value of the flux factor is a function of the planetary rotation rate as well as of the atmospheric mass. Planetary rotation rates are unlikely to vary greatly unless the planet is perturbed by another body of approximately planetary mass. The effect of variation in rotation rate has already been discussed in chapter 2. Here the rotation rate may have been considerably faster before lunar capture, and thus the value of the flux factor would be initially much higher. Model 2* includes this possible variation.

Model 2* ×

(i) *Figure A4* (*Upper graph*). $f = 4$ and $A = 0.07$, with the same incident flux as for Model 2, gives $T_e = T_s \simeq 260$ K ($\times 1$).

(ii) *Figure A2* (*Upper graph*). Low surface temperature leads to little degassing, but capture of the satellite modifies the value of f. Thus we have $T_s \simeq 290$ K ($\times 2$).

(iii) *Figure A3* (*Lower graph*). A higher value of T_s leads to rapid degassing and thus to increases in values of both the flux factor and the albedo. These increases combined with increasing solar flux lead to $T_s \simeq 315$ K ($\times 3$).

After this stage the evolutionary track of Model 2* would

probably follow that of Model 2—moving to the third circle and then onwards.

A further interesting discussion which is simplified by the use of isopleths of surface temperature presented in figures A1 to A5 is the consideration of a specific planetary feature, say, for instance, the ability of a planetary surface–atmosphere system to permit the origin and support the evolution of life. It is possible to suggest conditions under which the origin of life and its evolution would be likely. For the discussion illustrated here the conditions chosen are that the surface temperature shall be between 270 K and 320 K and that water vapour is in the atmosphere. These are reasonable conditions, but they may be over-restrictive (i.e. given these conditions for long enough, life stands a good chance of origin and evolution, but less hospitable conditions may not imply that life cannot exist, see §6.2). Zones corresponding to life-bearing conditions are considered for well established atmospheric systems (i.e. only for two of the graphs in figure A5). The areas calculated to lie within the 'hospitable' zone are shown in the dashed-line envelope. A number of interesting points arise from the illustration of these zones when considered with the evolutionary models described earlier. For instance, on the upper graph of figure A5 the fourth circle (which was the final evolutionary position of an early Earth-like planet—Model 2) lies within the life-bearing zone, whilst the second triangle (Venus-type runaway planetary evolution—Model 1) lies just outside it. A planet with an evolutionary history similar to that of Model 2 is a likely candidate for the origin and evolution of life (see §6.2) but a Model 1 type planet may also pass through a 'life-bearing' or at least a 'life-hospitable' phase. Reference to the lower graph of figure A5 shows that Model 1's (△3) final state is outside the region considered here as hospitable. If life became established on a planet which had characteristics typical of the upper graph, it would not find it difficult to adjust if the planetary system evolved to a state typical of the lower graph.

Bibliography

Abelson P H 1966 *Proc. Nat. Acad. Sci. USA* **55** 1365–72

Ahrens T J and O'Keefe J D 1972 *The Moon* **4** 214–49

Allen C W 1963 *Astrophysical quantities* 2nd edn (London: University of London Press) 263 pp

Alvarez L W, Alvarez W, Asaro F and Michel H V 1980 *Science* **208** 1095–108

Anders E and Owen T 1977 *Science* **198** 453–65

Arculus R J and Delano J W 1980 *Nature* **288** 72–4

Atreya S K, Donahue T M and Kuhn W R 1978 *Science* **201** 611–3

Bada J L and Miller S L 1968 *Science* **159** 423–5

Bahcall J N and Davis R Jr 1976 *Science* **191** 264–7

Barron E J, Sloan J L and Harrison C G A 1980 *Palaeogeog., Palaeoclim., Palaeoecol.* **30** 13–40

Barron E J, Thompson S L and Schneider S H 1981 *Science* **212** 501–8

Basu A R, Ray S L, Saha A K and Sarkar S N 1981 *Science* **212** 1502–6

Bayley S T and Morton R A 1979 in *Strategies of Microbial Life in Extreme Environments* ed M Shilo (Berlin: Dahlen Konferenzen) 109–24

Benlow A and Meadows A J 1977 *Astrophys. Space Sci.* **46** 293–300

Berger A 1977 *Nature* **269** 44–5

—— 1979 *Geophys. Surveys* **3** 351–402

—— 1980 *Vistas in Astronomy* **24** 103–22

—— 1981 *Astronomical theory of palaeoclimates in climatic variations and variability: facts and theories* ed A Berger (Dordrecht: Reidel) pp 501–25

Berkner L V and Marshall L C 1965 *J. Atmos. Sci.* **22** 225–61

Berkofsky L 1976 *J. Appl. Meteor.* **15** 1139–44

Bethoux J P 1978 in *Evolution of Planetary Atmospheres and Climatology of the Earth* (France: CNRS)

Birch F 1965 *Geol. Soc. Am. Bull.* **76** 133–53

Black D C and Suffolk G C J 1973 *Icarus,* **19** 353–7

Bolin V, Degens E T, Kempe S and Ketner P ed 1979 *The Global Carbon Cycle, SCOPE,* **13** (New York: Wiley)

Brancazio P J and Cameron A G W ed 1964 *The Origin and Evolution of Atmospheres and Oceans* (New York: Wiley)

Bray J R 1979 *Quaternary Res.* **12** 204–11

Broadfoot A L, Belton M J S, Takacs P Z, Sandel B R, Shemansky D E, Holberg J B, Ajello J M, Atreya S K, Donahue T M, Moos H W,

Bertaux J L, Blamont J E, Strobel D F, McConnell J C, Dalgano A, Goody R and McElroy M B 1979 *Science* **204** 979–82.

Broadfoot A L, Sandel B R, Shemansky D E, Holberg J B, Smith G R, Strobel D F, McConnell J C, Kumar S, Hunten D M, Atreya S K, Donahue T M, Moos H W, Bertaux J L, Blamont J E, Pomphrey R G, and Linick S 1981 *Science* **212** 206–11

Broecker W S 1981 *Glacial to interglacial changes in ocean chemistry* in *CIMAS Symposium (1980)* ed E Kraus (Miami: University of Miami)

Budyko M I 1969 *Tellus* **21** 611–9

Campbell C G and Vonder Haar T H 1980 *Atmos Sci. Paper 323* Colorado State University 74

Cameron A G W and Pollack J B 1976 *On the origin of the solar system and of Jupiter and its satellites* in *Jupiter* ed T Gehrels (Tucson: University of Arizona Press) pp 61–84

Canuto V and Hsieh S-H 1980 *Astrophys. J.* **237** 613–5

Carr M H 1980 *Space Sci. Rev.* **26** 231–85

Carr M H, Crumpler L S, Cutts J A, Greely R, Guest J E and Masursky H 1977 *J. Geophys. Res.* **82** 4055–65

Carr M H, Masursky H, Strom R G and Terrile R J 1979 *Nature* **280** 729–33

Cassen P, Reynolds R T and Peale S J 1979 *Geophys. Res. Lett.* **6** 731–4

Cess R D 1976 *J. Atmos. Sci.* **33** 1831–43

—— 1978 *J. Atmos. Sci.* **35** 1765–8

Cess R D, Briegleb B P and Lian M S 1982 *J. Atmos. Sci.* **39** 53–9

Cess R D, Ramanathan V and Owen T 1980 *Icarus* **41** 159–66

Chamberlain J W 1978 *Theory of Planetary Atmospheres: An Introduction to their Physics and Chemistry* (New York: Academic) 330 pp

Chameides W L, Walker J C G and Nagz A F 1979 *Nature* **280** 820–2

Chang S, DesMarias D, Mack R, Miller S L and Stratheam G 1981 Prebiotic organic synthesis and the origin of life in *Origin and Evolution of Earth's Earliest Biosphere: an Interdisciplinary Study* ed J W Schopf (Princeton, NJ: Princeton University Press) in press

Charney J G 1975 *Q. J. R. Met. Soc.* **101** 193–202

Christensen-Dalsgaard J and Gough D O 1976 *Nature* **259** 89–92

—— 1981 *Astron. Astrophys.* **104** 173–6

Christy J W and Harrington R W 1978 *Astron. J.* **83** 1007–8

Chylek P and Coakley J A Jr 1975 *J. Atmos. Sci.* **32** 675–9

Clark D H, McCrea W H and Stephenson F R 1977 *Nature* **265** 318–9

Cloud P E 1968 *Science* **160** 792–36

—— 1976 *Palaeobiology* **2** 351–87

Cloutier P A, McElroy M B and Michel F C 1969 *J. Geophys. Res.* **74** 6215–28

Cogley J G 1979 *Nature* **279** 712–3

—— 1981 personal communication on glacial events

—— 1982 *Mantle hydrosphere relationships of the early degassing of the Earth* private communication

Consolmagno G J and Lewis J S 1976 *Preliminary thermal history models of icy Galilean satellites in Jupiter* ed T Gehrels (Tuscon: University of Arizona Press) pp 1021–51

Corliss J B, Baross J A and Hoffman S E 1981 *Oceanologica Ada* **26** 59–69

Counselman C C M III, Gaurevitch S A, King R W, Loriat G B and Ginsberg E S 1980 *J. Geophys. Res.* **85** 8026–30

Crafoord C and Kallen E 1978 *J. Atmos. Sci.* **35** 1123–5

Cronin J R, Pizzarello S and Moore C B 1979 *Science* **206** 335–7

Cruikshank D P 1979 *Rev. Geophys. Space Phys.* **17** 165–76

Cruikshank D P and Silvaggio P H 1979 *Astrophys. J.* **233** 1016–20

Cutts J A, Blasius K R and Roberts W J 1979 *J. Geophys. Res.* **84** 2975–95

Dalfes H N, Thompson S L and Schneider S H 1979 *Bull. Am. Astron. Soc.* **11** 573 abs

Darius J 1975 *Cosmogony now* in *New Science in the Solar System* ed P Stubbs (London: IPC Magazines) 2–8

Dastoor M N, Margulis L and Nelson K H ed 1981 *Interaction of the Biota with the Atmosphere and Sediments* Final Report of the NASA Workshop in Global Ecology, NASA Headquarters, Washington, DC

Davies G F 1981 *Nature* **290** 208–13

Davis R 1979 *Status and Future of Solar Neutrino Research* vol 1, ed G Friedlander (Brookhaven: Brookhaven National Laboratory)

De Paulo D J 1981 *Trans. Am. Geophys. Union* **62** 137–40

Donahue T M, Hoffman J H and Hodges R R Jr 1981 *Geophys. Res. Lett.* **8** 513–6

Drobyshevskii E L M 1979 *Soviet Astron.* **23** 334–40

Ellis J S 1978 *Cloudiness, the Planetary Radiation and Climate, PhD Thesis* Colorado State University, Fort Collins

Enhalt D M 1974 *Tellus* **26** 58–70

Ewing M and Donn W L 1956 *Science* **152** 1706–12

Fanale F P 1971 *Icarus* **15** 279–303

Fanale F P, Brown R H, Cruikshank D P and Clarke R N 1979 *Nature* **280** 761–3

Fanale F P and Cannon W A 1979 *J. Geophys. Res.* **84** 8404–14

Farmer C B, Davies D W, Holland A L, LaPorte D D and Doms P E 1977 *J. Geophys. Res.* **82** 4225–48

Farmer C B and Doms P E 1979 *J. Geophys. Res.* **84** 2881–9

Ferris J P and Nicodem D E 1972 *Nature* **238** 268–9

Fink U, Benner D C and Dick K A 1977 *J. Quant. Spectrosc. Radiat. Transfer* **18** 447

Fink U and Larson H P 1979 *Astrophys. J.* **223** 1021–40

Fink U, Smith B A, Benner D C, Johnson J R and Reitsema H J 1980 *Icarus* **44** 62–71

Flint R F 1973 *The Earth and its History* (New York: W W Norton) 407 pp

Frakes L A 1979 *Climates Throughout Geologic Time* (Amsterdam: Elsevier) 310 pp

French R G and Gierasch P 1979 *J. Geophys. Res.* **84** 4634–42

Fricker P E and Reynolds R T 1968 *Icarus* **9** 221–30

Garrels R M and Mackenzie F T 1969 *Science* **163** 570–1

Gautier D and Courtin R 1979 *Icarus* **39** 28–45

Ghil M 1976 *J. Atmos. Sci.* **33** 576–601

Gierasch P J 1981 private communication, Goddard Institute for Space Studies (GISS) Summer School

Gierasch P J and Goody R M 1972 *J. Atmos. Sci.* **29** 400–2

Gierasch P J, Ingersoll A P and Pollard D 1979 *Icarus* **40** 205–12

Global Atmospheric Research Programme (GARP) 1975 *The physical basis of climate and climate modelling. Report of a study conference in Stockholm, 1974* GARP pubn No 16

Godson W L 1960 *Q. J. R. Met. Soc.* **86** 301–7

Goldring R 1972 in *Understanding the Earth* ed I G Gass, P J Smith and R C L Wilson (Sussex: Artemis) pp 157–61

Golitsyn, G S 1975 *Sov. Astron. Lett.* **1** 38–42

—— 1979 *Icarus* **38** 333–41

Goodwin A M 1981 *Science* **213** 55–61

Goody R M 1964 *Atmospheric Radiation: I. Theoretical Basis* (Oxford: Clarendon) 435 pp

—— 1975 *Weather on the inner planets* in *New Science in the solar system* ed P Stubbs (London: IPC Magazines) 39–42

Goody R M and Walker J C G 1972 *Atmospheres* (New York: Prentice-Hall) 150 pp

Grec G, Fossat E and Pomerantz M 1980 *Nature* **288** 541–4

Greenberg M 1982 Interstellar chemistry and comets, in *Cosmochemistry and the origins of life* ed C Ponnamperuma (Dordrecht: Reidel)

Grieve R A F and Dence M R 1979 *Icarus* **38** 230–42

Griffith E J, Ponnamperuma C and Gabel N W 1977 *Origins of Life* **8** 71–85

Guest J E 1980 Evolution of planetary surface (internal forces) in *The Evolution of terrestrial planets and major satellites* ed A J Meadows (Leicester: University of Leicester) pp 7–15

Guest J E, Butterworth P, Murray J and O'Donnell W 1979 *Planetary Geology* (London: David and Charles)

Gulkis S, Jansen M A and Olsen E T 1978 *Icarus* **34** 10–9

Hanel R, Conrath B, Flasar M, Herath L, Kunde V, Lowman P, Maguire W, Pearl J, Pirraglia J, Samuelson R, Gautier D, Gierasch P, Horn L, Kumar S and Ponnamperuma C 1979 *Science* **206** 952–6

Hanel R, Conrath B, Flasar F M, Kunde V, Maguire W, Pearl J, Pirraglia J, Samuelson R, Herath L, Allison M, Cruikshank D, Gautier D, Gierasch P, Horn L, Koppany R and Ponnamperuma C 1981 *Science* **212** 192–200

Hanks T C and Anderson D L 1969 *Phys. Earth Planet. Inter.* **2** 19–29

Hansen J E, Johnson D, Lacis A, Lebedeff S, Lee P, Rind D and Russell

G 1981 *Science* **213** 957–66

Hansen J E, Lacis A A, Lee P and Wang W-C 1980 *Climatic Effects of Atmospheric Aerosols* (New York: Academy of Sciences) 338 575–87

Hargrave R B 1976 *Science* **193** 363–71

Harrington R S and Van Flandern T C 1979 *Icarus* **39** 131–6

Hart M H 1974 *Icarus* **21** 242–7

—— 1978 *Icarus* **33** 23–39

Haselgrove C B and Hoyle F 1959 *Mon. Not. R. Astron. Soc.* **119** 112–3

Hays J D 1977 *Climate change and the possible influence of variations of solar output* in *The Solar Output and its Variations*, ed O R White (Boulder: Colorado Associated University Press) pp 73–90

Hays J D, Imbrie J and Shackleton N J 1976 *Science* **194** 1121

Head J W and Solomon S C 1981 *Science* **213** 62–76

Henderson-Sellers A 1976 *A Theoretical Study of the Evolution of the Atmospheres and Surface Temperatures of the Terrestrial Planets*, PhD Thesis University of Leicester

—— 1977 *Atmos. Environ.* **11** 864

—— 1978 *J. Geophys. Res.* **83** 5057–62

—— 1979 *Nature* **279** 786–8

—— 1981 *Geophys. Surveys* **4** 297–329

—— 1982 *The chemical composition and climatology of the Earth's early atmosphere* in *Cosmochemistry and the Origin of Life* ed C Ponnamperuma (Dordrecht: Reidel)

Henderson-Sellers A, Benlow A and Meadows A J 1980 *Q. J. R. Astron. Soc.* **21** 74–81

Henderson-Sellers A and Cogley J G 1982 *Nature* **298** 832–5

Henderson-Sellers A and Henderson-Sellers B 1975 *J. Atmos. Sci.* **32** 2358–60

—— 1980 *S. Afr. J. Sci.* **76** 511–3

Henderson-Sellers A and Hughes N A 1982 *Prog. Phys. Geog* **6** 1–45

Henderson-Sellers A and Meadows A J 1975 *Nature* **254** 44

—— 1976 *Planet. Space Sci.* **24** 41–4

—— 1977 *Nature* **270** 589–91

—— 1978 in *Evolution of Planetary Atmospheres and Climatology of the Earth* (Toulouse: Centre National D'Etudes Spatiales) pp 25–31

—— 1979a *Tellus* **31** 170–3

—— 1979b *Planet. Space Sci.* **27** 1095–9

Henderson-Sellers A and Schwartz A W 1980 *Nature* **287** 526–9

Henderson-Sellers B and Henderson-Sellers A 1978 *Nature* **272** 439–40

Hide R 1965 in *Magnetism and the Cosmos* ed W R Hindmarch, F J Lowes, P H Roberts and S K Runcorn (Edinburgh: Oliver and Boyd) pp 378–93

—— 1980 *The Evolution of Terrestrial Planets and Major Satellites* ed A J Meadows (Leicester: University of Leicester) pp 47–54

—— 1981 *Jupiter and Saturn: giant magnetic rotating fluid planets* Halley Lecture, Oxford University, *Observatory* **101**

Hoffert M I, Callegari A J, Hsieh C T and Ziegler W 1981 *Icarus* **47** 112–29

Hoffman J H and Hodges R R 1975 *The Moon* **14** 159–67

Hoffman J H, Oyama V I and Von Zahn U 1980 *J. Geophys. Res.* **85** 7871–90

Holland H D 1962 in *Petrologic Studies: A volume in honour of A F Buddington Geol. Soc. Am.* pp 447–79

—— 1972 *Geochim. Cosmochim. Acta* **36** 637–51

—— 1978 *The Chemistry of the Atmosphere and Oceans* (New York: Wiley Interscience) 351 pp

Horowitz N H, Hobby G L and Hubbard J S 1977 *J. Geophys. Res.* **82** 4659–62

Howard J N, Burch D E and Williams D 1956 *J. Opt. Soc. Am.* **46** 186–90, 237–41, 242–5, 334–8

Hoyle F and Wickramasinghe C 1979 *Diseases from Space* (London: Dent)

Hughes D W 1980 *New Sci.* **85** 66–9

Hunten D M 1973 *J. Atmos. Sci.* **30** 1481–94

—— 1977 Titan's atmosphere and surface in *Planetary Satellites* ed J A Burns (Tucson: University of Arizona Press) pp 420–37

Imhoff C I and Giampapa M S 1980 *Astrophys. J.* **239** L115–9

Ingersoll A P 1969 *J. Atmos. Sci.* **26** 1191–8

—— 1981 *Sci. Am.* **245** 90–108

Irvine W M, Leschine S B and Schloerb F P 1981 *Origin of Life* ed Y Wolman (Dordrecht: Reidel) pp 27–32

Jakosky B M and Ahrens T J 1979 *Proc. 10th Lunar Sci. Conf.* (New York: Pergamon) p 3043

Johnson J R, Fink U, Smith B A and Reitsena H J 1981 *Icarus* **46** 288–91

Kasting J F 1979 *Evolution of Oxygen and Ozone in the Earth's Atmosphere, PhD Thesis* University of Michigan

—— 1982 *J. Geophys. Res.* **87** 3091–8

Kasting J F and Donahue T M 1980 *J. Geophys. Res.* **85** 3255–63

Kasting J F and Walker J C G 1981 *J. Geophys. Res.* **86** 1147–58

Keating G M, Nicholson J Y III and Lake L R 1980 *J. Geophys. Res.* **85** 7941–56

Keeling C D, Adams J A, Ekdahl C A and Guenther P R 1976a *Tellus* **28** 552–64

Keeling C D, Bacastow R B, Bainbridge A E, Ekdahl C A, Guenther P R and Waterman L S 1976b *Tellus* **28** 538–51

Knauth L P and Epstein S 1976 *Geochim. Cosmochim. Acta* **40** 1095–108

Kreifeledt J G 1971 *Icarus* **14** 419–30

Kuhn W R and Atreya S K 1979 *Icarus* **37** 207–13

Kuhn W R, Rodgers S R and MacElroy R D 1979 *Icarus* **37** 336–46

Kumar S 1976 *Icarus* **28** 579–91

Kvenvolden K A, Lawless J, Pering K, Peterson E, Flores J, Ponnamperuma C, Kaplan I R and Moore C 1970 *Nature* **228** 923–6

Lahav N, White D H and Chang S 1978 *Science* **201** 67–9

Lamb H H 1970 Volcanic dust in the atmosphere; with a chronology and assessment of its meteorological significance. *Phil. Trans. R. Soc.* A **266** 425–533

Lazcano-Araujo A R and Oro J 1981 *Cometary Material and the Origins of Life on Earth* in *Comets and the Origin of Life: Proceedings of the Fifth College Park Colloquium on Chemical Evolution* ed C Ponnamperuma and D B D Donn

Lebofsky L A, Rieke G H and Lebofsky M J 1979 *Icarus* **37** 554–8

Leighton R B and Murray B C 1966 *Science* **153** 136–44

Leovy C B and Zwek R W 1979 *J. Geophys. Res.* **84** 2956–68

Levine J S 1978 *Comparative Planetology*, ed C Ponnamperuma (New York: Academic) 165–82

—— 1980 *Origins of Life* **10** 313–23

Levine J S and Allario F 1982 *Environ. Monitor. Assess.* **1** 263–306

Levine J S and Augustsson T R 1981 *Did the Earth ever have a Jovian Prebiological Paleoatmosphere?* in *Proc. 6th College Park Colloquium on Chemical Evolution, University of Maryland, College Park, MD, 5–6 October 1981*

Levine J S, Augustsson T R, Boughner R E, Natarajan M and Sacks L 1981 *Comets and the Origin of Life* ed C Ponnamperuma (Dordrecht: Reidel) pp 161–90

Levine J S, Augustsson T R and Hoell J M 1980a *Geophys. Res. Lett.* **7** 317–20

Levine J S and Boughner R E 1979 *Icarus* **39** 310–4

Levine J S, Boughner R E and Smith K A 1980b *Origins of Life* **10** 199–213

Levine J S, Hays P B and Walker J C G 1979a *Icarus* **39**

Levine J S, Hughes R E, Chameides W L and Howell W E 1979b *Geophys. Res. Lett.* **6** 557–9

Levine J S, Kraemer D R and Kuhn W R 1977 *Icarus* **31** 136–45

Levine J S, McDougal D S, Anderson D E and Barker E S 1978 *Science* **200** 1048–51

Levine J S and Schryer D R 1978 *Man's Impact on the Troposphere*, NASA, Langley Publications 1022 pp

Lewis J S 1974 *Sci. Am.* **230** (3) 50–66

Li Y-H 1972 *Am. J. Sci.* **272** 119–37

Lian M S and Cess R D 1977 *J. Atmos. Sci.* **34** 1058–62

Lockwood G W and Thompson D T 1979 *Nature* **280** 43–5

Lovelock J E 1975 *Proc. R. Soc.* B **189** 167

—— 1979 *Gaia, A New Look at Life on Earth* (Oxford: Oxford University Press) 157 pp

Lucchita B K, Anderson D M and Shoji H 1981 *Nature* **290** 759–63

Lynden-Bell D and Pringle J E 1974 *Mon. Not. R. Astron. Soc.* **168** 603–37

McCrea W H 1975 *Nature* **255** 607–9

—— 1981 *Proc. R. Soc.* A **375** 1–41

McCulloch M T. Gregory R T, Wasserburg G J, Taylor H P Jr 1981 *J. Geophys. Res.* **86,** 2721–36

Macdonald G J F 1964 *Escape of helium from the Earth's atmosphere* in *The origin and evolution of atmospheres and oceans* ed P J Brancazio and A G W Cameron (New York: Wiley) pp 127–82

McDonough T R and Brice N M 1973 *Icarus* **20** 136–45

McElhinny M W 1973 *Palaeomagnetism and plate tectonics* (Cambridge: Cambridge University Press) 358 pp

—— 1979 *The Earth: its Origin, Structure and Evolution* (New York: Academic) 597 pp

McElhinny M W and Burek P J 1971 *Nature* **232** 98–102

McElroy M B 1972 *Science* **175** 443–5

—— 1975 *The atmosphere and ionosphere of Jupiter* in *Atmosphere of the Earth and Planets* ed B M McCormac (Dordrecht: Reidel) pp 409–23

—— 1976 in *Chemical Kinetics* vol 9, ed D R Herschbach (London: Butterworths)

McElroy M B, Kong T Y and Yung Y L 1977 *J. Geophys. Res.* **82** 4379–88

McElroy M B and Prather M J 1981 *Nature* **293** 535–9

McElroy M B, Yung Y L and Nier A O 1976 *Science* **194** 70

McGill G E 1979 *Geophys. Res. Lett.* **6** 739–41

Macy W Jr 1979 *Icarus* **40** 213–22

Manabe S and Wetherald R T 1967 *J. Atmos. Sci.* **24** 241–59

Margulis L and Lovelock J E 1974 *Icarus* **21** 471–89

Margulis L, Walker J C G and Rambler M 1976 *Nature* **264** 620–4

Martin T Z 1975 *PhD Thesis* University of Hawaii, Honolulu

Mason B 1958 *Principles of Geochemistry* (New York: Wiley)

Masursky H, Boyce J M, Dial A L, Schaber G G and Strobell M E 1977 *J. Geophys. Res.* **82** 4016–38

Matheja J and Degens E T 1971 *Structural Molecular Biology of Phosphates* (Stuttgart: Gustav Fischer) 180 pp

Meadows A J 1972 *Phys. Letts.* **5C** 199–235

Miller S L and Orgel L E 1974 *The Origins of Life on Earth* (Princeton NJ: Prentice-Hall) 229 pp

Mills A A 1980 *J. Br. Astron. Ass.* **90** 219–30

Morabito L A, Synnott S P, Kupferman P N and Collins S A 1979 *Science* **204** 972

Moorbath S 1982 *The dating of the earliest sediments on Earth* in *Cosmochemistry and the Origins of Life* ed C Ponnamperuma (Dordrecht: Reidel)

Moorbath S, O'Nions R K and Pankhurst R J 1973 *Nature* **245** 138–9

Moorbath S and Windley B F 1981 *The Origin and Evolution of the Earth's Continental Crust* (London: Royal Society) 303 pp

Murray B C, Ward W R and Yeung S C 1973 *Science* **180** 638–40

Mutch T A 1979 *Rev. Geophys. Space Phys.* **17** 1694–755

Namias J 1978 *Mon. Weath. Rev.* **100** 893–7

Ness N F, Acuna M H, Lepping R P, Connerney J E P, Behannon K W, Burlaga L F and Neobauer F M 1981 *Science* **212** 211–7

Newman M J and Rood R T 1977 *Science* **198** 1035–7

North G R 1975 *J. Atmos. Sci.* **32** 2033–43

Oort A H and Vonder Haar T H 1976 *J. Phys. Oceanograph.* **6** 781–800

Oparin A I 1938 *The Origin of Life* (New York: Dover) 267 pp

Otterman J 1975 *Science* **189** 1013–5

—— 1977 *Climatic Change* **1** 137–57

Owen T 1976 *Icarus* **28** 171–7

—— 1978 in *Evolution of Planetary Atmospheres and Climatology of the Earth* (Toulouse: Centre Nationale D'Etudes Spatiales) pp 1–10

Owen T, Cess R D and Ramanathan V 1979 *Nature* **277** 640–2

Ozima M, 1975 *Geochim. Cosmochim. Acta* **39** 1127–34

Parmentier E M and Head J W 1979 *J. Geophys. Res.* **84** 6263–76

Pearl J, Hanel R, Kunde V, Maguire W, Fox K, Gupta S, Ponnamperuma C and Raulin F 1979 *Nature* **280** 755–8

Pettengill G H, Campbell D B and Masursky H 1980 *Sci. Am.* **242** (8) 54–65

Pilcher C B 1979 *Icarus* **37** 559–74

Pilcher C B, Morgan J S and Macy W W 1979 *Icarus* **39** 54–64

Podolak M and Bar-Nun A 1979 *Icarus* **39** 272–6

Podolak M and Cameron A G W 1974 *Icarus* **22** 123–48

Pollack J B 1971 *Icarus* **14** 295–306

—— 1979 *Icarus* **37** 479–553

Pollack J B and Black D C 1979 *Science* **205** 56–9

Pollack J B, Burns J A and Tauber M E 1979 *Icarus* **37** 587–611

Pollack J B, Grossman A S, Moore R and Graboske H C Jr 1977 *Icarus* **30** 111–28

Pollack J B, Rages K, Toon O B and Yung Y L 1980a, *Geophys. Res. Lett.* **7** 829–33

Pollack J B, Toon O B and Boese R 1980b *J. Geophys. Res.* **85** 829–33

Pollack J B and Yung Y L 1980 *Ann. Rev. Earth Planet Sci.* **8** 425–87

Ponnamperuma C ed 1982 *Cosmochemistry and the Origins of Life* (Dordrecht: Reidel)

Randall D 1982 *Banded iron formations in the primitive terrestrial environment* private communication

Rambler M B and Margulis L 1980 *Science* **210** 638–40

Rampino M R, Self S and Fairbridge R W 1979 *Science* **206** 826–9

Ransford G A, Finnesty A A and Collerson K D 1981 *Nature* **289** 21–4

Rasool S I and De Bergh C 1970 *Nature* **226** 1037–9

Redfield A C 1958 *Am. Sci.* **46** 205–21

Rind D 1981 private communication, Goddard Institute for Space Studies

Ringwood A E 1979 *The Origin of the Earth and Moon* (New York: Springer) 295 pp

Ripley E A 1976 *Science* **191** 100

Roads J O 1978 *J. Atmos. Sci.* **35** 753–73

Ronov A B and Yaroshevsky A A 1967 *Geochim Int.* 1041–66

Rossow W B 1978 *Icarus* **36** 1–50

Rossow W B, Del Genio A D, Limaye S S, Travis L D and Stone P H 1980 *J. Geophys. Res.* **85** 8107–29

Rubey W W 1951 *Geol. Soc. Am. Bull.* **62** 1111–48

Rycroft M J 1982 *Nature* **295** 190–1

Sagan C 1975a *Sci. Am.* **233** (3) 23–31

—— 1975b *J. Atmos. Sci.* **32** 1079–83

Sagan C and Mullen G 1972 *Science* **177** 52–6

Sagan C, Toon O B and Gierasch P J 1973 *Science* **181** 1045–9

Sanchez R A, Ferris J P and Orgel L E 1966 *Science* **154** 784–5

—— 1967 *J. Mol. Biol.* **30** 223–53

Scarf F L, Taylor W W L, Russell C T and Brace L H 1980 *J. Geophys. Res.* **85** 8158–66

Schidlowski M 1980a *Antiquity of photosynthesis: possible constraints from Archean carbon isotope record* in *Biogeochemistry of Ancient and Modern Environments,* ed P A Trudinger and M R Walters (Canberra: Australian Academy of Science) pp 47–54

—— 1980b *The atmosphere* in *The Handbook of Environmental Chemistry* vol 1, Part A, ed O Hutzinger Berlin: Springer) pp 1–16

Schidlowski M, Appel P W U, Eichman R and Junge C E 1979 *Geochim. Cosmochim. Acta* **43** 189–99

Schneider S H 1972 *J. Atmos. Sci.* **29** 1413–22

—— 1975 *J. Atmos. Sci.* **32** 2060–6

Schneider S H and Mass C 1975 *Science* **190** 741–6

Schneider S H and Thompson S L 1980 *Icarus* **41** 456–70

Schneider S H, Washington W H and Chervin R M 1978 *J. Atmos. Sci.,* **35** 2207–21

Schopf J W 1980 private communication, University of Nijmegen

Schwarzschild M, Howard R and Harm R 1957 *Astrophys. J.* **125** 233–41

—— 1958 *Structure and Evolution of Stars* (Princeton, NJ: Princeton University Press) 295 pp

Schwartz A W ed 1972 *Theory and Experiment in Exobiology* (Netherlands: Wolters-Noordhoff)

—— 1981 *Chemical Evolution—the Genesis of the First Organic Compounds in Marine Organic Chemistry* eds E K Dieursna and R Dawson (Amsterdam: Elsevier) pp 7–30

Seiff A and Kirk D B 1977 *J. Geophys. Res.* **82** 4364–78

Seiff A, Kirk D B, Young R E, Blanchard R C, Findlay J T, Kelly G M and Sommer S C 1980 *J. Geophys. Res.* **85** 7903–33

Sellers W D 1969 *J. Appl. Meteor.* **8** 392–400

—— 1976 *Mon. Weath. Rev.* **104** 233–48

Seyfert C K and Sirkin L A 1979 *Earth History and Plate Tectonics,* 2nd edn (New York: Harper and Row) 504 pp

Shklovskii I S and Sagan C ed 1966 *Intelligent Life in the Universe* (New York: Holden Day)

Slobodkin L S, Buyakov I F, Triput N S, Caldwell J and Cess R D 1981 *J. Quant. Spectrosc. Radiat. Transfer* **26** 33–8

Slobodkin L S, Buyakov I F, Triput N S, Cess R D, Caldwell J and Owen T 1980 *Nature* **285** 211–3

Smagorinski J 1981 in *Climatic Variations and Variability: Facts and Theories* ed A Berger (Dordrecht: Reidel) pp 661–87

Smith B A, Soderblom L A, Beebe R, Boyce J, Briggs G, Bunker A, Collins S A, Hansen C J, Johnson T V, Mitchell J L, Terrile R J, Carr M, Cook A F II, Cuzzi J, Pollack J B, Danielson G E, Ingersoll A, Davies M E, Hunt G E, Masursky H, Shoemaker E, Morrison D, Owen T, Sagan C, Veverka J, Strom R, Suomi V E 1981 *Science* **212** 163–90

Smith B A, Soderblom L A, Johnson T V, Ingersoll A P, Collins S A, Shoemaker E M, Hunt G E, Masursky H, Carr M H, Davies M E, Cook A F, Boyce J, Danielson G E, Owen T, Sagan C, Beebe R F, Veverka J, Strom R, McCauley J F, Morrison D, Briggs A G, Suomi V E 1979 *Science* **204** 951–71

Smoluchowski R 1967 *Nature* **215** 691–5

Soffen G A 1977 *J. Geophys. Res.* **82** 3959–70

Spitzer C R ed 1980 *Viking Orbiter Views of Mars* (Washington DC: NASA)

Stacey F D 1980 *Phys. Earth Planet. Inter.* **22** 89–96

Stoks P G and Schwartz A W 1981 *Geochim. Cosmochim. Acta* **45** 563–9

Strobel D 1981 *Chemistry and the Evolution of Titan's Atmosphere* in *Proc. College Panel Symp. on the Voyager Mission: Implications for Planetary Biology* ed C Ponnamperuma

Strobel D F and Yung Y L 1979 *Icarus* **37** 256–63

Sutton J and Windley B F 1974 *Sci. Prog.* **61** 401–20

Taylor F W 1981 private communication regarding reworking of the net flux measurements made by Pioneer Venus

Taylor F W, Beer R, Chahine M T, Diner D J, Elson L S, Haskins R D, McCleese D J, Matonchik J V, Reichley P E, Bradley S P, Delderfield J, Schofield J T, Farmer C B, Froidevauz L, Leung J, Coffey M T and Gille J C 1980 *J. Geophys. Res.* **85** 7963–8006

Thomas P, Veverka J and Campos-Marquetti R 1979 *J. Geophys. Res.* **84** 4621–33

Toksoz M N, Hsui A T and Johnston D H 1978 *The Moon and the Planets* **18** 281–320

Tokunaga A, Knacke R F and Owen T 1975 *Astrophys. J. Lett.* **197** L77

Tomasko M G, Smith P H, Suomi V E, Sromovsky L A, Revercomb H E, Taylor F W, Martonchik D J, Seiff A, Boese R, Pollack J B, Ingersoll A P, Schubert G and Covey C C 1980 *J. Geophys. Res.* **85** 8187–99

Toupance G, Mourey D and Raulin F 1978 in *Evolution of planetary atmospheres and climatology of the Earth* (Toulouse: Centre Nationale d'Etudes Spatiales) pp 31–40

Toupance G, Raulin F and Buvet R 1975 *Origins of Life* **6** 83–90

Trafton L 1977 *Icarus* **31** 369–84

—— 1981 *Rev. Geophys. Space Phys.* **19** 43–89

Tyler G L, Eshleman V R, Anderson J D, Levy G S, Lindal G F, Wood G E and Croft T A 1981 *Science* **212** 201–6

Waldrop M M 1981 *Science* **211** 470–2

Walker J and Rowntree P R 1977 *Q. J. R. Met. Soc.* **103** 29–46

Walker J C G 1975 *J. Atmos. Sci.* **32** 1248–56

—— 1976 *Implications for atmospheric evolution of the inhomogeneous-accretion model for the origin of the Earth* in *In the early history of the Earth*, ed B F Windley (New York: Wiley) pp 537–46

—— 1977 *Evolution of the Atmosphere* (New York: Macmillan) p 318

—— 1978a *Atmospheric Constraints on the Evolution of Metabolism* presented at *Fourth College Park Symposium on Chemical Evolution, University of Maryland, 18–20 October 1978*

—— 1978b *Pageoph.* **116** 222–31

—— 1978c *Comparative Planetology* 141–63

—— 1979 *Pageoph.* **117** 498–512

—— 1982 *Paleogeog., Paleoclim., Paleoecol* submitted

Walker J C G, Hays P B and Kasting J F 1981 *J. Geophys. Res.* **86** 9776–83

Wallace D and Sagan C 1979 *Icarus* **39** 385–400

Wang W-C, Rossow W B, Yao M-S and Wolfson M 1981 *J. Atmos. Sci.* 1167–78

Ward W R 1973 *Science* **181** 260–2

Warren S G and Schneider S H 1979 *J. Atmos. Sci.* **36** 1377–91

Weare B C and Snell F M 1974 *J. Atmos. Sci.* **31** 1725–34

Wetherald R T and Manabe S 1980 *J. Atmos. Sci.* **37** 1485–510

Wetherill G W 1975 *Ann. Rev. Nucl. Sci.* **25** 283–328

—— 1980 *Ann. Rev. Astron. Astrophys.* **18** 77–113

Wick G L 1971 *Science* **173** 1011–2

Wigley T M L and Brimblecombe P 1981 *Nature* **291** 213–5

Williams G P and Robinson J B 1973 *J. Atmos. Sci.* **30** 684–717

Williams I P 1975 *The Origin of the Planets* (Bristol: Adam Hilger)

Wilson L 1980 *Evolution of planetary and satellite surfaces (external forces)* in *The Evolution of Terrestrial Planets and Major Satellites* ed A J Meadows (Leicester: University of Leicester) pp 15–21

Windley B F ed 1976 *The Early History of the Earth* (London: Wiley) 619 pp

—— 1977 *The Evolving Continents* (New York: Wiley) 385 pp

—— 1980 in *The Evolution of Terrestrial Planets and Major Satellites* ed A J Meadows (Leicester: University of Leicester) pp 3–7

Winston J S, Gruber A, Gray T I, Varnadove M S, Earnest C L and Mannello L P 1979 *Earth–Atmosphere Radiation Budget Analyses Derived from NOAA Satellite Data, June 1974–February 1978* vol 1–2 (Washington, DC: NOAA/NESS)

Wolfendale A W 1978 *New Sci.* **79** 634–6

Yung Y L and Pinto J P 1978 *Nature* **273** 730–2

Subject Index

229

230

Hydrodynamical flow, 42
Hydrogen, 12–4, 23, 24, 40, 42–3,
    45–7, 52, 53, 62, 63, 74, 79, 80,
    83, 84–5, 99–100, 105, 118–20,
    123, 132, 134, 140, 143, 147–50,
    151–3, 191, 193
Hydrogen chloride, 163
Hydrogen cyanide, 118, 126, 145
Hydrogen sulphide, 86, 143
Hydrosphere, 13, 28, 29, 61, 94, 104,
    110, 112, 125, 128, 130, 134–5,
    138–9, 149–50, 156–8, 172, 175,
    187, 196, 200, 202, 205, 206
Hydrostatics, 39

ITC, 176, 178
Iapetus, 96, 97, 101, 165
Ice, 96, 97, 100–1 see also Volatiles,
    condensation
Impact/bombardment (craters), 25–7,
    65, 94, 100–1, 110, 126–8, 131–3,
    135, 162, 170, 171, 185, 198, 200,
    203
Impact/vaporisation, 28, 136
Infrared absorption, 21, 49–59, 65, 67,
    83–6, 88–9, 95, 109, 117, 118,
    121, 124, 142–3, 144, 163, 190
Inhomogeneous accretion model, 6, 7
Intelligence, 202
Internal energy, gravitational, 7, 33,
    37, 39, 73–9, 80–2, 86, 100, 114,
    116, 130, 171
Internal feedback processes, 22, 33
Interstellar
    dust, 161, 169, 197, 200
    gases, 2
Io, 2, 8, 10, 11, 43, 44, 96, 97, 98–100,
    118, 123, 162, 171
Ionosphere, 48
Iridium, 127
Ishtar, 110
Isothermal atmosphere, 39
Isua, 87, 132, 139

Jean's escape, 42, 47, 61, 99
Jet stream (Earth), 35, 80–2
Jovian atmospheres, 56 §3.1
Jovian (major) planets, 9, 21, 35, 47,
    73–81, 82–6, 172, 201
Jupiter, 2–5, 8, 10, 11, 20, 25, 35–8, 40,

43, 44, 48, 56, 71, 73–85, 86, 96,
    97, 98–101, 117, 123, 202
Jupiter's satellites, 7, 98–9, 101

Kaolinite, 104
Krypton, 12, 23, 24, 46, 52, 53, 105,
    106, 107, 108

Laminated terrain (Mars), 38
Lapse rate, 41, 48, 55, 56, 67, 70, 83,
    90, 109, 117, 129, 140, 150, 181,
    196, 198
Latent heat, 19, 55, 68, 125, 197
Lava flows, 186
Life, 72, 90, 101, 112, 125–6, 129–30,
    132–3, 149–50, 154–6, 158, 195–6
    adaptation to extreme conditions,
    201
Lightning, 53, 73, 110, 132, 144, 149,
    170, 203
Linear features, 96, 97, 100, 110
Lithosphere, 27, 113, 116–7, 130, 158,
    168, 187

Magnetic field (Jupiter), 43, 85, 99
Magnetic fields (planetary), 43, 85,
    93–4, 99, 168, 170
Main sequence, 15
Mantle, 13, 25, 100, 108, 130, 133–6,
    165
Mariner, 93, 104, 105, 114, 120, 166,
    185, 186
Mars, 2–5, 8, 10, 11, 25, 28, 34–8, 42,
    48, 51, 56, 59, 63, 65–6, 72, 91,
    92, 94, 101–7, 108, 110, 111, 114,
    115, 116–7, 120, 124, 135, 144,
    149, 160–6, 168, 172, 174–5, 182,
    183, 184, 185–7, 188, 189–90,
    192–3, 196, 198, 201, 206
    polar cap, 38, 65, 117, 163, 167,
    172, 174, 183, 184, 189, 190, 206
    subsurface ice, 65
Mauna Loa, 62, 157
Maxwell (Venus), 110
Maxwell–Boltzmann distribution, 47
Mercury, 2–5, 8, 10, 11, 24, 34, 48, 56,
    72, 92–5, 101, 108, 111, 165
Meteorite (see also carbonaceous
    chondrite), 9, 14, 25, 30, 79–80,
    106, 107, 108, 126–7, 133, 135,

232

Author Index

237

238